Civil Engineering and Engineering Mechanics

Civil Engineering and Engineering Mechanics

Finn Gibson

NY RESEARCH PRESS
New York

Published by NY Research Press
118-35 Queens Blvd., Suite 400,
Forest Hills, NY 11375, USA
www.nyresearchpress.com

Civil Engineering and Engineering Mechanics
Finn Gibson

International Standard Book Number: 978-1-64725-421-6 (Hardback)

Cataloging-in-Publication Data

Civil engineering and engineering mechanics / Finn Gibson.
p. cm.
Includes bibliographical references and index.
ISBN 978-1-64725-421-6
1. Civil engineering. 2. Mechanics, Applied. 3. Engineering mathematics.
4. Engineering. I. Gibson, Finn.

TA153 .C58 2023
624--dc23

Contents

Preface

Over the recent decade, advancements and applications have progressed exponentially. This has led to the increased interest in this field and projects are being conducted to enhance knowledge. The main objective of this book is to present some of the critical challenges and provide insights into possible solutions. This book will answer the varied questions that arise in the field and also provide an increased scope for furthering studies.

Civil engineering refers to a professional engineering discipline concerned with the design, building and maintenance of physical and naturally created environment including dams, roads, railways, sewage systems, bridges, airports, canals, pipelines, buildings and their structural components. Civil engineering is traditionally divided into several sub-disciplines. It is related with the knowledge of materials science, structures, geology, geography, hydrology, soils, mechanics, environmental science and project management. Engineering mechanics refers to the use of mechanics to address real-world problems using fundamental engineering components. It involves the study of micro and nanomechanics, dynamics and vibrations of complex structures, mechanics of biological and energy systems, mechanical properties of materials, aero and hydrodynamics, and stress analysis of fixed and movable structures. This book provides significant information of engineering mechanics to help develop a good understanding of civil engineering. It strives to provide a fair idea about this discipline and to help develop a better understanding of the latest advances within this field.

I hope that this book, with its visionary approach, will be a valuable addition and will promote interest among readers. Each of the authors has provided their extraordinary competence in their specific fields by providing different perspectives as they come from diverse nations and regions. I thank them for their contributions.

Finn Gibson

1

Building Materials and Construction

1.1 Building Material

Building material is any material which is used for a construction purpose. Many naturally occurring substances, like clay, sand, wood and rocks, even twigs and leaves have been used to construct buildings. Apart from naturally occurring materials, several man-made products are in use, some more and some less synthetic. The manufacture of building materials is an established industry in several countries and the use of these materials is usually segmented into specific specialty trades, like carpentry, plumbing, roofing and insulation work.

1.1.1 Stones

Stone is an important building material and stones are economical to build houses especially in the hilly areas. Stones have been used since ancient times as building material. Roads are paved with stone. Monuments such as Tajmahal were built using stones. Many temples and palaces are made from stones. Stones are more long lasting than other building materials. Stone is a good choice for foundations in areas which are likely to be flooded during rains. Submerged bricks may breakdown but submerged stones remain stable. Today stones form an important source of aggregates and are also used in ornamental work.

A good building stone should have the following qualities:

- Appearance: In face work the stone should have fine, compact texture. Light-Coloured stone is preferred as dark colours are likely to fade out in due course of time.
- Structure: A broken stone should not be dull in appearance and should have uniform texture. It should free from cavities, cracks and patches of loose or soft material. Stratification should not be visible to naked eye.
- Strength: A stone should be strong and durable to withstand the disintegrating action of weather. Compressive strength of building stones in range between 60 to 200 N/mm2.
- Weight: Weight is an indication of the porosity and density. For stability of

structures such as dams, retaining walls etc. heavier stones are required whereas for arches, vaults, domes, etc. light stones may be the choice.

- Hardness: This property is important for floors, pavements, aprons of bridges, etc. The hardness is determined by the Mohr scale.
- Toughness: The measure of impact is that a stone can withstand is defined as toughness. The stone used should be tough when vibratory otherwise moving loads are anticipated.
- Porosity and Absorption Porosity depends on the mineral constituents, cooling time and structural formation. A porous stone disintegrates as the absorbed rain water freezes, expands and causes cracking.
- Seasoning: The stone should well-seasoned.
- Weathering: The resistance of stone against the wear and tear due to natural agencies must be high.
- Workability: Stone should be workable. The cutting, dressing and bringing the stone in the required shape and size may not be uneconomical.
- Fire Resistance: Stones should be free from calcium carbonate, oxides of iron and minerals having different coefficients of thermal expansion. Igneous rock shows marked disintegration principally because of quartz which disintegrates into small particles at a temperature of about 575°C. Limestone can withstand a little higher temperature up to 800°C after which they disintegrate.
- Specific Gravity: The specific gravity of most of the stones lies between 2.3 to 2.5.
- Thermal Movement: Thermal movements alone are usually not trouble-some. However joints in coping and parapets open-out in letting the rain water causing trouble. Marble slabs show a distinct distortion when subjected to heat. An exposure of one side of marble slab to heat may cause that side to expand and the slab warps. On cooling the slab does not go back to its original shape.

Criteria for Selection

The following are the various characteristics to be considered while selecting stones:

1. Strength: Factors affecting strength are:

- Hardness: The composition of the compounds determines its hardness. Stones with Silica and Sodium are soft while with magnesium and Iron are hard.
- Size and shape of the minerals in stones: Crystalline solids are harder and

compact compared to non-crystalline solids. Finer the crystals stronger are the stones as it reduces porosity.

- Cohesion (binding force): It is the property of the particles to attract each other. Fine grains are more cohesive than coarse grains. Cohesion directly affects stone's strength.
- Density and porosity: If a stone is dense specific gravity above 2.7 it is strong. High porosity means less density. Porosity reduces stone's resistance to concentrated (point) load.

2. Durability: The stones must last long so they must have greater resistance to weathering, extreme temperatures, abrasion etc. Stones for road metal, floor slab, etc. have to endure abrasion. To increase the durability, quarrying and cutting of stone must be done with great care. Polished stones are more durable as water slides off easily from the former.

3. Resistance to heat: The stone must have a very low amount of expansion due to increase in temperature. Limestone resists fire better than stones containing quartz as quartz explodes on heating.

4. Appearance: The aesthetic aspects of stone, (i.e.) colour, look, texture etc. must be considered for selection.

5. Cost: Cost of the stone is influenced by quality of the stone, transportation charges, mode of purchase and difficulties in quarrying and cutting.

Tests on Stones

- Absorption test**:** Take a sample of stone. Immerse it in distilled water for a day. Remove the stones from water and weigh to know the water absorbed. It should not exceed 0. 6%.
- Smith's test: To find out the presence of soluble matter in stone take few pieces of stone and place them in a glass tube filled with clear water. After an hour shake the tube. Clear water indicates that the stone is durable and free from any soluble matter and vice versa.
- Acid test: Take a sample of stone. Place it in a 5% H_2SO_4 or HCL solution for some days. Then take out the stones and dry them. Broken edges and powdery surface indicates the presence of Calcium Carbonate ($CaCO_3$) which has poor weathering quality. This test is usually carried out on sandstone.
- Freezing and thawing test: This test is done to find out the frost resisting properties of a stone. Immerse stones in freezing water at place it at a temperature of–12°C for 24 hours. Take the stones out and keep them at room temperature.

Observe the changes on their surface. If the changes are less the stones are freeze resistance.

- Crystallization test: Immerse stones in 14% sodium sulfate solution for 2 hours and dry it in an oven at 1000C. Repeat the procedure for 5 times. Check for loss of weight or cracks on the stone. If the changes are minimal then the stones have passed the test. Crystallization test is prescribed by Bureau of Indian Standards (BIS).

- Hardness test: The Dorry's testing machine is used to check the hardness of stones. Take a cylindrical stone of diameter and height each 25mm and place in the machine at a pressure of 12.5N. Rotate the machine at a speed of 28 rpm. After 1000 revolutions take and weigh the stone. The coefficient of the hardness is determined by the following calculation.

$$\text{coefficient of hardness} = 20 - \left(\frac{\text{loss in weight in grams}}{3}\right)$$

If coefficient of hardness is greater than 17 the stone is hard. If it is between 14 and 17 the stone of medium hardness and if it is less than 14 stone is of poor hardness.

- Impact test: This test is done to determine the toughness of the stone with the "impact machine". Take a cylinder of diameter and height each 25 mm from the stone sample. Place it on cast iron anvil of the machine. Allow a steel hammer of weight 20N to fall axially over the cylinder. Height of first blow is 1cm that of second blow is 2cm and so on.

- Microscopic test: Place thin sections of stone under microscope. Examine grain size, constituents, binding materials and presence of harmful materials.

- Attrition test: To find out the rate of wear or abrasion of stone due to grinding action when vehicles run over it deval's attrition testing machine is used.

  Break the stones into 60 mm sizes and fill 50g of broken pieces into the cylinders of Deval's machine. Rotate the cylinders at a rate of 30 rpm for 5 hours. After 5 hours take the pieces out and pass them through a sieve of size 1.5 mm mesh. Weigh the pieces of stones retained on the sieve and calculate the loss in weight.

$$\%\text{ of wear} = \frac{\text{Loss in weight}}{\text{initial weight}} \times 100$$

- Crushing test: To determine compressive strength of stones compression testing machine is used. Cover three stones cubes of side 10 cm with plaster of Paris. Subject them to load in compression testing machine one by one and note the values of loads at which the blocks get crushed. Find the average crushing load. Let it be P.

$$\text{Therefore ultimate crushing strength} = \frac{P}{\text{base area of stone}}$$

Here base area = 10 cm × 10 cm = 100cm².

For a good building stone crushing strength should be greater than 100N/mm².

1.1.2 Bricks

Brick is one of the oldest building material and continues to be a most popular and leading construction material because of being cheap, durable and easy to handle and work with. For building-up exterior and interior walls, partitions, piers, footings and other load bearing structures clay bricks are used.

A brick is rectangular in shape and of size that can be conveniently handled with one hand. Brick may be made of burnt clay or mixture of sand and lime or of Portland cement concrete. Clay bricks are commonly used because they are economical and easily available.

The length, width and height of a bricks are interrelated as below:

Length of brick = 2 × width of brick + thickness of mortar.

Height of brick = Width of brick.

Size of a standard brick should be 19 × 9 × 9 cm and 19 × 9 × 4 cm.

When placed in masonry the 19 × 9 × 9 cm brick with mortar becomes 20 × 10 × 10 cm. However the bricks available in most part of the country are 9" × 5" × 3".These are known as field bricks. Weight of such a brick is 3.0 kg. An indent called frog 1–2 cm deep as shown in figure is provided for 9 cm high bricks. The size of frog should be 10 × 4 × 1 cm. The purpose of providing frog is to form a key for holding the mortar and therefore the bricks are laid with frogs on top. Frog is not provided in 4 cm high bricks and extruded bricks.

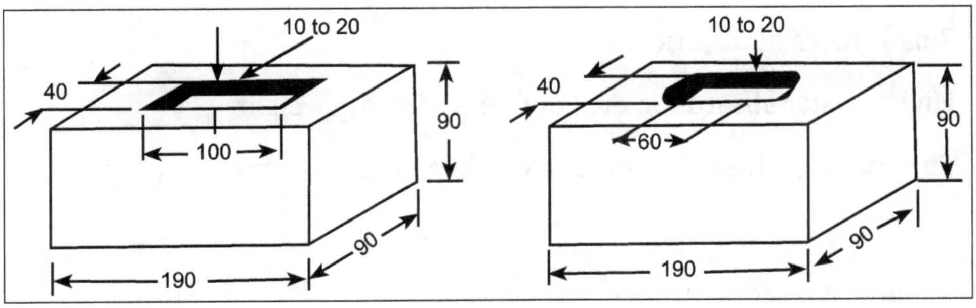

Bricks Classification

First Class Bricks

The properties of first class bricks are as follows:

- Well burnt and in deep red color.
- Smooth and rectangular surface with parallel straight and sharp edges.
- Free from cracks.
- Uniform texture.
- No impression should be left on the brick when a scratch is made by a finger nail.
- A metallic or ringing sound should come when two bricks are struck against each other. Water absorption should be around 15% of its dry weight when immersed in cold water for 24 hours.
- Crushing strength should not be less than 10 N/mm^2.

Uses

- It is used in painting.
- It is exposed face work in masonry.
- It is used in structural work.
- It is used in flooring.
- It is used in reinforced brick work.

Second class bricks:

- The properties of second class bricks are:
- Uniformly coloured and slightly over-burnt bricks.
- Slightly distorted surface and a little rounded edges.
- Small cracks in the bricks.
- Higher water absorption of about 20% of its dry weight.
- The crushing strength should not be less than 7 N/mm^2.

Uses:

- All hidden masonry works.

- Centering of reinforced brick and reinforced cement concrete structures.

Third Class Bricks

Following are the properties of third class bricks:

- Slightly under-burnt or over-burnt bricks.
- Soft and distorted with rounded edges.
- Rough and uneven surface.
- Dull sound produced when two bricks are struck against each other.
- Water absorption is about 25% of its dry weight.

Uses:

- To build temporary structures.

Fourth Class Bricks

- The properties of fourth class bricks are as follows:
- Over burnt in dark color.
- Badly distorted in shape and size.
- Brittle in nature.

Uses:

- It is used in aggregates for concrete in foundations.
- It is used for flooring in lime concrete.
- It is used in road metal.

On the Basis of Use

- Common bricks: These are ordinary multipurpose and economical bricks without any special appearance. The strength and durability of these bricks vary greatly. They are used in filling, backing and in wall where appearance is not important.
- Facing bricks: They are primarily made for aesthetic purpose with good colour and texture. They are highly durable under adverse climatic conditions. These bricks are used in facing or exterior construction for their pleasing appearance.
- Engineering bricks: These bricks are strong, impermeable, smooth, table

moulded, hard and conform to defined limits of absorption and strength. These type of bricks are used for all important load bearing structures.

On the Basis of Finish

- Sand-faced Brick has textured surface manufactured by sprinkling sand on the inner surfaces of the mould.
- Rustic Brick has mechanically textured finish varying in pattern.

On the Basis of Manufacture

- Hand-made: These bricks are hand moulded.
- Machine-made: Depending upon mechanical arrangement bricks are known as wire-cut bricks. Bricks cut from clay extruded in a column and cut off into brick sizes by wires are pressed bricks. When bricks are manufactured from stiff plastic or semi dry clay and pressed into moulds. Bricks are moulded by machines imitating hand mixing.

On the Basis of Burning

- Pale Bricks are under burnt bricks obtained from outer portion of the kiln.
- Body Bricks are well burnt bricks occupying central portion of the kiln.
- Arch Bricks are over burnt also known as clinker bricks obtained from inner portion of the kiln.

On the Basis of Types

- Solid: Small holes not exceeding 25 per cent of the volume of the brick are permitted alternatively frogs not exceeding 20 per cent of the total volume are permitted.
- Perforated: Small holes may exceed 25 per cent of the total volume of the brick.
- Hollow: The total of holes which need not be small may exceed 25 per cent of the volume of the brick.
- Cellular: Holes closed at one end exceed 20 per cent of the volume.

nufacturing of Bricks

ollowing are the steps involved in brick manufacture.

aration of clay - It involves the following steps:

Unsoiling: The top 20 cm layer of soil is taken and thrown away to remove impurities.

- Digging: The clay is dug out from the ground and spread on the surface.
- Cleaning: The clay is cleaned of plants etc. Clay lumps are powdered with crushing roller.
- Weathering: The clay is exposed to atmosphere for weeks or a full season for softening.
- Blending: The weathered clay is blended after adding sufficient ingredient by turning small portion of up and down every time.
- Tempering: The clay is made amply hard and homogeneous by kneading or pressing with water.

2. Moulding: Moulding is done manually or with machines.

- Hand moulding: Done on small scale where manpower is cheap with wooden or steel rectangular moulds open at the top and bottom. Hand-moulded bricks are divided into:
    - Ground moulded bricks: On leveled ground fine sand is sprinkled. Tempered clay is pressed into a mould dipped in water removing extra clay with strike. The mould is lifted to get raw brick. A mark is fixed showing manufacturer's trade name in the frog.
    - Table moulded bricks: Clay is filled in a mould and covered with a thin board called pallet. The mould is lifted off the stock board with frog projection nailed on moulding table so that it rests on the pallet. The mould is then removed to get raw bricks.
- Machine moulding: Machine moulded bricks are divided into:
    - Wire cut bricks: As clay is extruded out of a rectangular opening it is cut into strips by wires frames. The opening and strips are equal to the brick's dimensions. The bricks are called wire cut bricks.
    - Pressed bricks: Strong clay is converted to powder. Little water is added to form a stiff plastic paste, moulded and pressed by machine to form hard bricks known as pressed bricks. These bricks do not require drying.

Machine made bricks are more uniform and well-shaped than handmade bricks.

3. Drying: Drying saves time and fuel and prevents cracking of bricks while burning. Bricks are dried in open air driers or in artificial driers. The artificial dryers are of two types, hot floor drier and tunnel drier.

- Open air drier or Natural process: For a small scale of manufacturing freshly moulded bricks are laid on an even clean ground under sheds.

- Artificial drying process: For a large scale manufacturing hot floor drier method is used where bricks are filled, dried and emptied in rotation. Tunnel drier method enables bricks to be loaded at one end and subjected to heat inside and the dried bricks are taken at the other end.

4. Burning: It gives hardness, density, durability and strength to bricks. Over-burnt bricks are brittle and under burnt ones are soft. Burning is carried out either in clamps or kilns.

- Clamps: Clamps are temporary structures. The bricks and fuel are placed in alternate layers. The fuel amount is reduced from bottom to top. Each brick tier consists of 4-5 layers. For free heat circulation space is left between bricks. This process takes many days for burning bricks.
- Kiln: A kiln is a large oven. This method is fast and used to dry large quantity of bricks. There are of two types namely intermittent kiln and continuous Kiln.

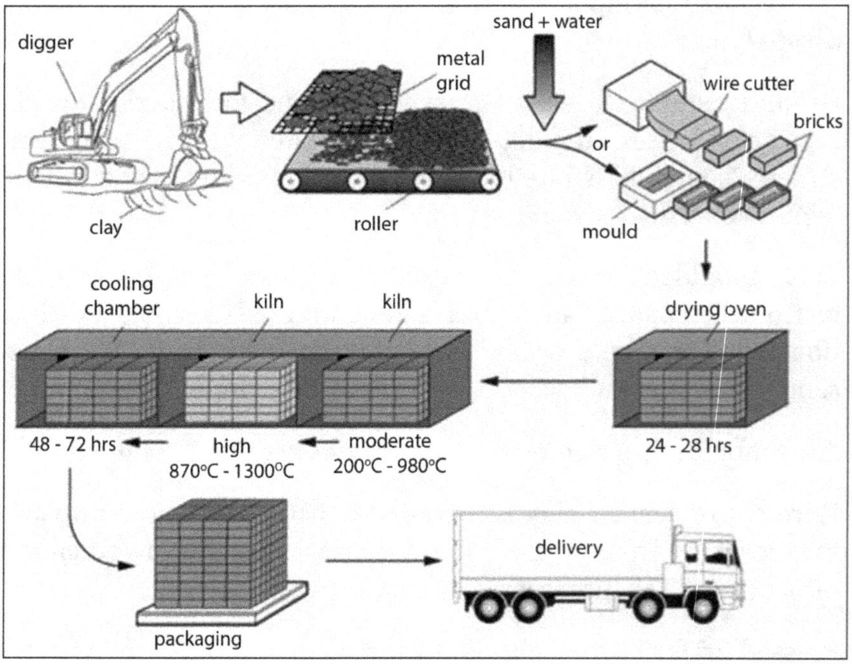

Bricks manufacturing process.

Test on Bricks

1. Compressive Strength

The Compressive strength is the capacity of a material or structure to withstand loads tending to reduce size. It can be measured by plotting applied force against deformation in a testing machine. Some materials fracture at their compressive strength limit others deform irreversibly so a given amount of deformation may be considered as the limit for compressive load.

Compressive strength is a key value for design of structures. Five bricks are taken at random and their dimensions are measured to 1 mm accuracy. They are then immersed in water of 25°C to 29°C for 24 hours.

The surplus moisture is allowed to drain and the frog if any is filled with mortar 1:3 (1 cement, 3 clean coarse sand 3 mm and down). It is kept under a jute bag for another 24 hours after which it is immersed in clean water for three days.

At the time of testing these bricks are removed from water, wiped dry of any trace of moisture and placed with the flat surface horizontal and mortar filled face up between three plywood sheets each of 3 mm thickness (plaster of Paris may also be used to ensure uniform surface).

The load is applied at the rate of 140 kg/cm2 per minute till the failure of the specimen takes place as indicated by the needle of the testing machine turning back. Average of the five test values is reported. While finding the average any single value obtained as compressive strength which is higher than the upper value of the class of the bricks tested should be taken only as the upper limit of the class. Values less than 20% below the average value should be discarded. The average value should not be less than the specified value.

2. Water Absorption

Five bricks are taken for test. They are allowed to dry in an oven at 110 to 115° C till they attain a constant weight which usually takes place in 48 hours. They are then allowed to cool at room temperature which generally takes 4 to 6 hours without a fan and 2 to 3 hours with a fan blowing on it and weight W_1 is measured.

They are then kept in clear water at 27 + 2°C for 24 hours and then wiped dry with a damp cloth and weight W) is measured. The average percentage of water absorbed as percentage of dry weight is reported. Average of the five tests is reported. This value should not be more than the values specified in table.

$$\text{Percentage absorption} = W_2 - W_1 \times 100.$$

Immerse 3 blocks in clean water at room temperature for one day. Then take them out and drain them for 1 minute and weigh them. Mark the weight as A. After weighing, dry them in oven at 100 to 115°C for 1 day and weigh them again. Note it as B. Calculate the absorption as below:

Where,

A = Wet mass of unit in Kg.

B = Dry mass of unit in Kg.

The water absorption should not be more than 10% of the mass.

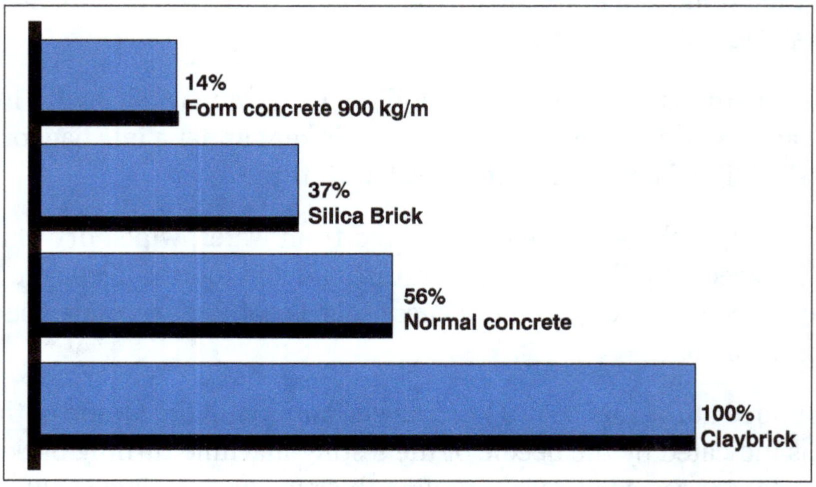

Water absorption test.

3. Efflorescence of Bricks

It is the powdery substance found on bricks and rocks surface due to weathering. Place a brick vertically in a vessel filled with water of 25mm depth. Allow the whole water to be absorbed by the brick. Repeat the procedure again. Presence of Efflorescence is classified as below:

- Nil: Efflorescence is not visible.
- Slight: When Efflorescence covers below 10% of exposed area of the brick.
- Moderate: When Efflorescence is above 10% and below 50% of exposed area of the brick.
- Heavy: When Efflorescence is above 50% but does not powder away the brick surface.
- Serious: When the deposits are heavy and flake away the brick surface.

This test should be conducted in a well ventilated room at 18-30°C.Average value on five samples taken at random is to be reported. The brick is placed vertically in a dish 30 cm x 20 cm approximately in size with 2.5 cm immersed in distilled water. The whole water is allowed to be absorbed by the brick and evaporated through it.

After the bricks appear dry a similar quantity of water is placed in the dish and the water is allowed to evaporate as before. The brick is to be examined after the second evaporation and reported as follows:

- Nil: When there is no perceptible deposit of salt.

- Slight- When not more than 10 per cent of the area of brick is covered with salt.
- Moderate- When there is heavy deposit covering up to 50% of the area of the brick but unaccompanied by powdering or flaking of the surface.
- Heavy- When there is heavy deposit covering more than 50% of the area of the brick accompanied by powdering or flaking of the surface.
- Serious-When there is a heavy deposit of salts accompanied by powdering and flaking of the surface and this deposition tends to increase in the repeated wetting of the specimen.
- Bricks for general construction should not have more than slight to moderate efflorescence. Efflorescence should not be more than moderate.

Efflorescence

4. Dimensional Tolerance

Select 20 bricks at random to check their length, width and height. Variations in dimensions are allowed only within narrow limits ± 3% for class I bricks, other classes ± 8%.Twenty whole bricks are selected at random to check measurement of length, width, height, etc. These dimensions are to be measured in one or two lots of ten each as shown in figure. Variations in dimensions are allowed only within narrow limits ±3 % for class one and ±8% for other classes.

5. Hardness

Scratch the brick's surface with your finger nail. If no impression is left on the surface the brick is hard. A scratch is made on the surface of the brick with the finger nail. In a good brick no impression will be left on the surface. In this test a scratch is made on brick surface with a hard thing. If that doesn't left any impression on brick then that is good quality brick.

6. Soundness

Take 2 bricks and strike them with each other. The bricks should not break and a clear ringing sound should be produced. In this test two bricks are held by both hands and struck with one another. If the bricks give clear metallic ringing sound and don't break then those are good quality bricks.

7. Structure

A brick is broken and its structure is examined. It should be homogeneous, compact and free from any defects such as holes, lumps etc. In this test a brick is broken or a broken brick is collected and closely observed. If there are any flows, cracks or holes present on that broken face then that is not good quality brick.

1.1.3 Cement

A building material is made by grinding calcined limestone and clay to a fine powder which can be mixed with water and poured to set as a solid mass or used as an ingredient in the making of mortar or concrete.

Cement is a binder. A substance that sets and hardens and can bind other materials together. Cements used in construction can be characterized as being either hydraulic or non-hydraulic depending upon the ability of the cement to be used in the presence of water. Non-hydraulic cement will not set in wet conditions or underwater rather it sets when it dries and reacts with carbon dioxide in the air. It can be attacked by some aggressive chemicals after the setting.

Hydraulic cement is made by replacing some of the cement in a mix with activated aluminium silicates, pozzolanas such as fly ash. The chemical reaction results in hydrates that are not very water soluble and so are quite durable in water and it is safe from chemical attack. This allows setting in wet condition or underwater and further protects the hardened material from chemical attack (e.g., Portland cement).

Use

- It is used in cement mortar for masonry work, plaster and pointing etc.
- It is used in concrete for laying floors, roofs and constructing lintels, beams, weather-shed, stairs, pillars etc.
- It is used in construction for important engineering structures such as bridge, culverts, dams, tunnels, light house, clocks, etc.
- It is used in construction of water wells, tennis courts, septic tanks, lamp posts, telephone cabins etc.
- It is used in making joint for joints, pipes, etc.
- It is used in manufacturing of precast pipes, garden seats, artistically designed wens, flower posts, etc.

- It is used in the preparation of foundation, water tight floors, footpaths, etc.

Ingredients

The concrete consisting of cement, sand and coarse aggregates mixed in a suitable proportions in addition to water is called cement concrete. In this type of concrete cement is used as a binding material, sand as fine aggregates and gravel, crushed stones as coarse aggregates.

The Advantages of Adding Pozzolana in Cement

- Economy: Industrial wastes such as fly ash are cheap hence by adding them cement cost can be reduced.
- Reduces permeability of concrete: As pozzolana is fine it fills the pores of concrete fully. All the free lime present in cement reacts with pozzolana forming a dense mass.

Manufacturing Process

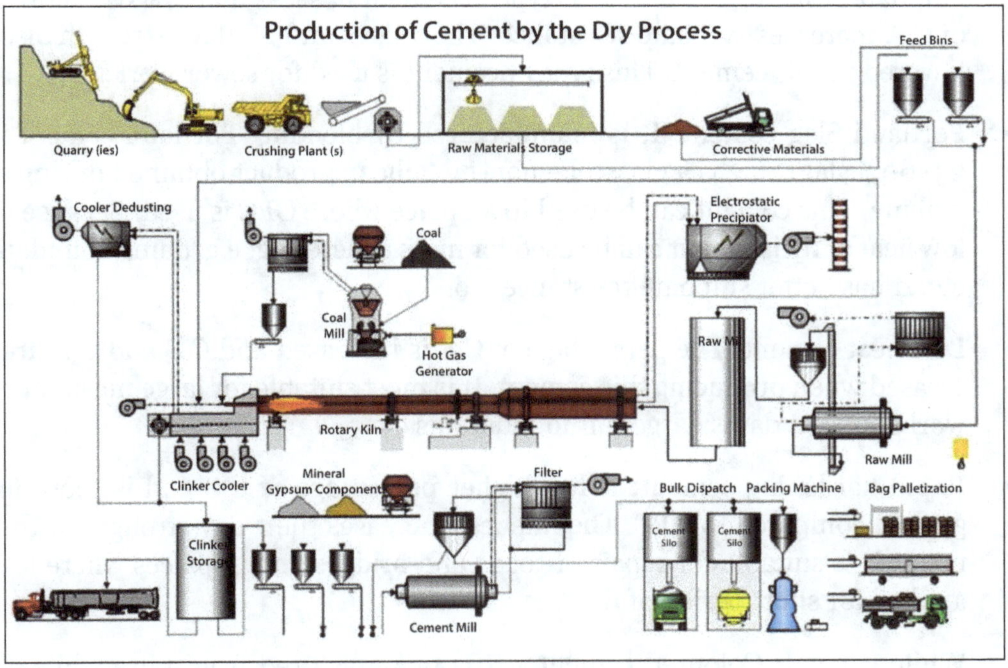

Manufacturing of cement by wet process.

Advantages of Wet Process

- Low cost of excavating and grinding raw material.
- Accurate control of composition and homogeneity of the slurry.

- Economical utilization of fuel through the elimination of separated drying operations.

Disadvantages of Wet Process

- Longer kiln are used in wet process, which cost is more.
- Longer kilns are less responsive to variable clinker demand than the short kilns which can be used in dry process.

Types and Grades

- Ordinary Portland Cement (OPC): It is the most common type of cement in general use. It is available in 3 grades in India.
- Portland Pozzolana Cement (PPC): This type of cement is very common in the market and is available in 3 grades. It is made by adding 10 to 25% pozzolana (fly ash or burnt clay) with OPC.
- Sulphate resisting cement: This type of cement has less than 5% C_3A. Lowering of C_3A increases sulphate resistance. The 7 days and 28 days strength also get lowered in this cement. This type of cement is used for sewer works.
- Portland Slag cement: It is manufactured by blending Portland cement with 25- 60% slag. Slag is a glass-like non-metallic byproduct obtained during steel making. This cement can be used in all place where OPC is used. Because of its low heat of hydration it can be used for mass concreting. e.g. dams, foundations etc. It has better sulphate resistance too.
- Low heat cement: The percentage of C_2S is increased and C_3S and C_3A are decreased when producing this cement. It is most suitable for large mass concrete works such as dams, large raft foundations etc.
- Rapid hardening cement: It has higher percentage of C_3S and is more finely ground compared to OPC. The fineness increases the 7 day strength of the cement. It is suitable for repair of roads and bridges and in places where load is applied for short period of time.
- White cement: Coloured cement - It is manufactured from pure white chalk and clay that is free from iron oxide. Iron oxide causes Grey colour in cement. Due to this the iron oxide content is limited below 1 percent. Coloured cements are made by adding 5 to 10 percent colouring pigments before grinding. These cements have same properties as that of OPC. Coloured cements are used for making terrazzo flooring, face plaster of walls (stucco), ornamental works and - casting stones.

- Hydrophobic cement: It is also called water repellent cement. A small amount of (0.1 to 0.5%) hydrophobic substances such as stearic acid, boric acid or oleic acid is mixed with ordinary Portland cement during grinding of clinker. A water repellent film is formed around cement particles during manufacturing itself. This prevents the entry of atmospheric moisture hence setting of cement during storage is prevented. During mixing with aggregates this film is broken and the cement behaves like ordinary cement. It is most suitable for basements and for making water tight concrete.

- High alumina cement: This is not a type of Portland cement and is manufactured by blending 40 percent bauxite, 40 percent lime, 15% iron oxide with a little of ferric oxide and silica magnesia etc. at a very high temperature. The alumina content should not be less than 32%. The resultant product is ground finely. Since C_3A is not present the cement has good sulphate resistance. It is not quick setting. It is resistant to action of fire, sea water, acidic water and sulphates and is used as refractory concrete in industries and is used for pre-casting.

- Super sulphated cement: It is manufactured by a mixture of granulated blast furnace slag not less than 70% Calcium sulphate and small quantity of 33 grade Portland cement. In this cement C_3A which is prone to sulphates is limited to less than 3.5%. It has low heat of hydration and is resistant to chemical attacks and in particular to sulphates.

- Water proof cement: It is manufactured by adding streates of Calcium (Ca) and Aluminium (Al) and gypsum treated with tannic acid at the time of grinding. It is resistant to penetration of water. Water proof cement is used in water retaining structures like tanks, reservoirs, retaining walls, swimming pools, bridge piers etc.

- Quick setting cement: The quantity of gypsum is reduced and small percentage of aluminum sulphate is added. It is ground much finer than OPC. The initial setting time is 5 minutes and the final setting time is 30 minutes. It is used when concrete is to be laid under water.

The following are commonly used artificial pozzolanas:

- Fly ash.
- Surki.
- Blast furnace slag.
- Rice husk ash.
- Silica fume.

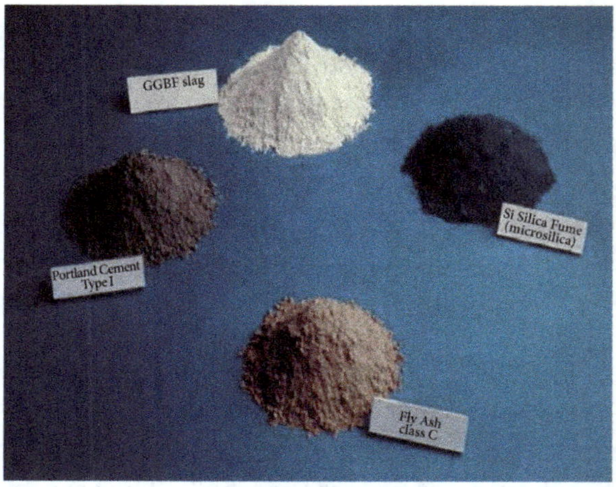

Artificial pozzolanas.

When mixed with ordinary Portland cement the silica of the pozzolana combines with the free lime released during the hydration of cement. This action is called pozzolanic action. The pozzolanic activity is due to the presence of fine glassy silica and lime which produces calcium silicate hydrate.

The silica in the pozzolana reacts with the lime produced during hydration of OPC and contributes to development of strength. Slowly and gradually additional calcium silicate hydrate is formed which is a binder and fills up the space, gives impermeability, durability and strength.

Hydration of Portland cement may be expressed as:

$$C_3S + H_2O \rightarrow C - S - H \rightarrow + Ca(OH)_2$$

(Calcium silicate hydrate) (Lime).

Lime produced combines with silica of pozzolana:

$$Ca(OH)_2 + SiO_2 \rightarrow C - S - H$$

(Silica)

The optimum amount of pozzolana as replacement for cement may normally range between 10-30% for artificial pozzolanas. It may be as low as 4-6% for natural pozzolanas.

Grades

The OPC has been classified into three grades, namely 33 grade, 43 grade and 53 grade based upon the strength of the cement. If the 28 days strength is not less than 33 N/mm2, it is called 33 grade cement. If the strength is not less than 43 N/mm^2, it is called 43 grade cement. If the strength is not less than 53 N/mm^2, it is called 53 grade cement.

Type of cement	Fineness	Soundness mm	Setting Time		28 Days compressive strength MPa
			Initial min	Final min	
33 Grade OPC	225	10	30	600	33
43 Grade OPC	225	10	30	600	43
53 Grade OPC	225	10	30	600	53

Properties of Cement and Cement Mortar

- Setting time.
- Soundness.
- Fineness.
- Strength.

The cement found in mortar provides strength and fills in air pockets present in sand. Sand on the other hand eliminates or reduces cracks produced by the contraction of cement while it sets. In most cases adding one part cement to two parts sand will yield a proportional cement mortar volume. Cement mortar also contains Pozzuoli admixture an active mineral which is a combination of silica fume (SF) and low calcium content fly ash (FA). These elements contribute to the flexural strength and elasticity of cement mortar.

Test for Setting Time

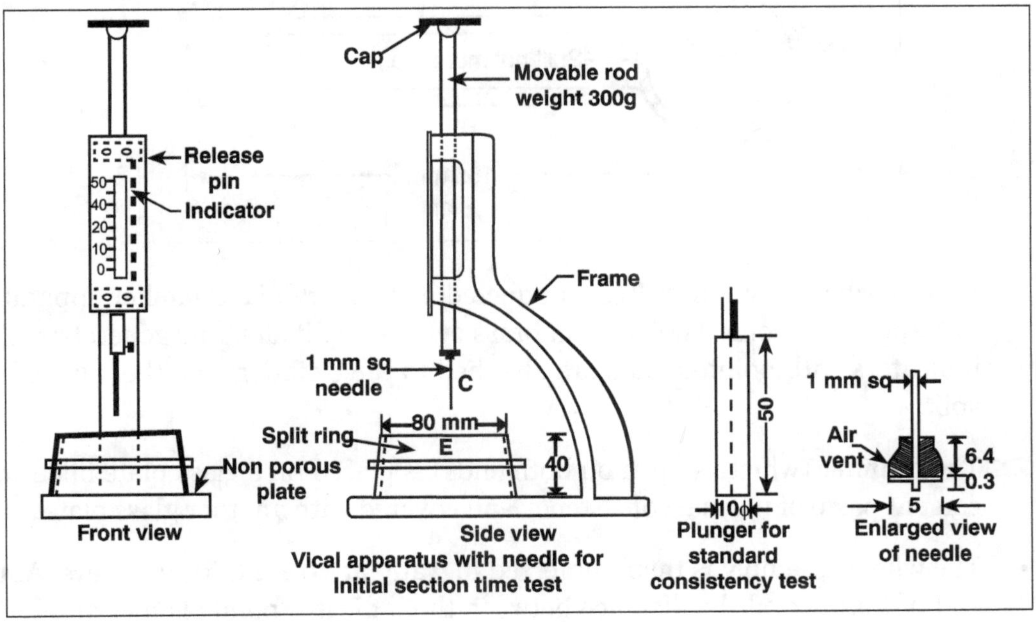

- The setting time test is conducted by using the same Vicat apparatus, except that a 1mm diameter needle is used for penetration.
- The test is started about 15 minutes after placing the cement paste (which has normal consistency) into the mold. Trials for penetration of the needle are made.
- The final setting time is defined as the length of time between the penetration of the paste and the time when the needle (with annular ring) no longer sinks visibly into the paste.
- The initial setting time is defined as the length of time between the penetration of the paste and the time when the needle penetrates 25 mm into the cement paste.

Test for Soundness

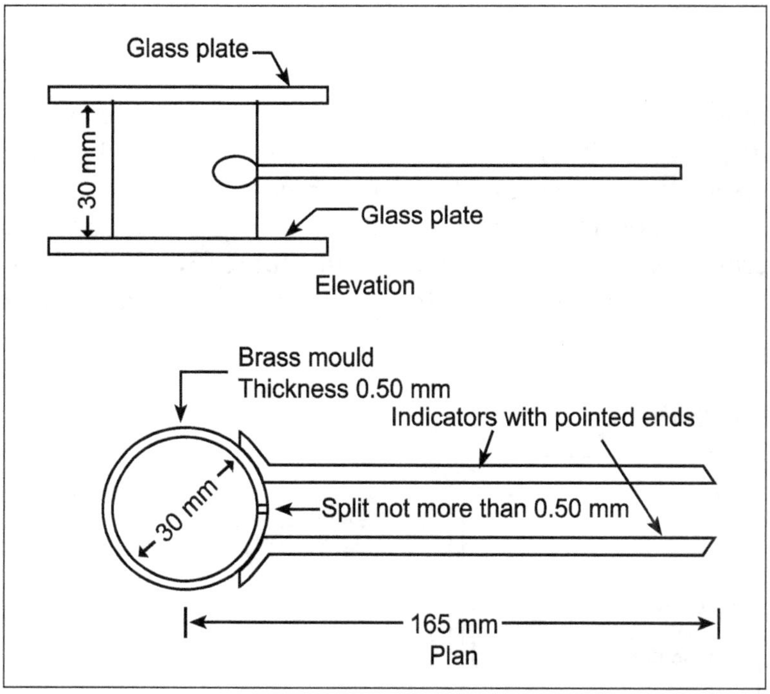

- IS prescribe a Soundness Test conducted by using the Le Chatelier apparatus. The apparatus consists of a small brass cylinder split along its generatrix. Two indicators with pointed ends are attached to the cylinder on either side of the split.
- The cylinder (which is open on both ends) is placed on a glass plate filled with cement paste of normal consistency and covered with another glass plate.
- The whole assembly is then immersed in water at 20 ± 1°C for 24 hours. At the end of that period the distance between the indicator points is measured. The

mold is then immersed in water again and brought to a boil. After boiling for one hour the mold is removed from the water, after cooling, the distance between the indicator points is measured again. This increase represents the expansion of the cement paste for Portland cements, expansion is limited to 10mm.

Test for Fineness

- There are various methods for determining the fineness of cement particles. The Blaine air-permeability method is the most commonly used method.
- In the Blaine air-permeability method, given volume of air is passed through a prepared sample of definite density. The number and size of the pores in a sample of given density is a function of the particles and their size distribution and determines the rate of air flow through the sample. Calculations are made and the fineness is expressed in terms of cm^2 /g or m^2 /kg.

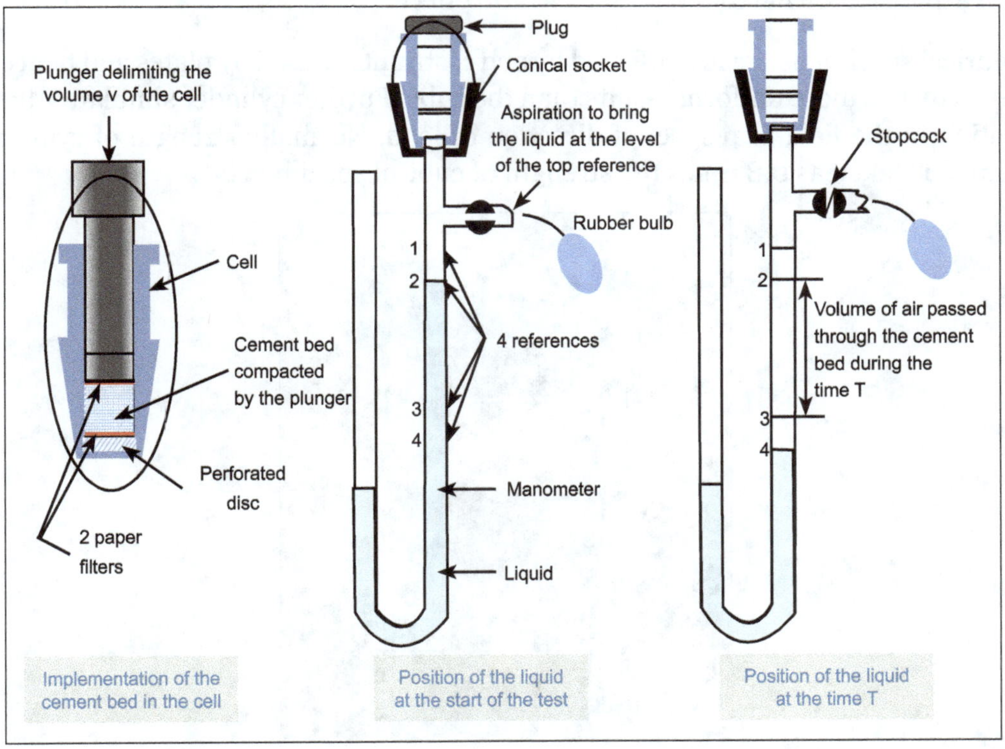

Compressive Strength Test

This is an important test. Most of the properties of concrete are qualitatively related to it. It is an easy and most common test. The tests are conducted on cubical or cylindrical specimens. The cube specimen is of size 15 × 15 × 15 cm and the cylinder is about 15 cm diameter and 30 cm long.

The largest nominal size of the aggregates does not exceed 20 mm. The moulds are to be of metal moulds, preferably of steel or cast iron. The moulds are made in such a way that the specimen is taken out without damage. A tamping steel bar of 16 mm diameter 0.6 m long with a bullet end is used for compacting.

The test cube specimens are made as soon as practicable. The concrete is filled into the mould in 5 m deep approximately. Each layer is compacted by tamping rod (25 to 35 strokes depending on 10 or 15 cm depth) or by vibration. After the top layer has been compacted the top of the mould is leveled using a trowel. The top is covered with a glass or metal plate to prevent evaporation.

The specimens are de -moulded after 24 hours and submerged in fresh, clean water or saturated lime solution and kept there until taken out just prior to testing. The water should be maintained approximately at 27° C ± 2° C and on no account the specimens are allowed to become dry. The specimen is tested in a compression testing machine at the completion of 7 days and 28 days. Compression on the cube or cylinder undergoes lateral expansion owing to the Poisson's ratio effect.

Cylindrical specimens are less affected by end restraints caused by plates and hence it is believed to give more uniform results than the cube. Further cylinder simulates the real condition in the field in respect of direction of load. Normally strength of cylindrical specimen is taken as 0.8 times the strength of cubical specimens.

Field Testing

The Following are the Field Tests for Cement that are carried out in order to ascertain the quality of Cement:

- Colour: The colour of cement must be uniform. It must be typical cement colour i.e., grey colour with a light greenish shade. This is not always a reliable test. But it gives an indication of excess lime or clay and the degree of burning.

- Physical properties: The cement must feel smooth when touched or rubbed in between fingers. If it is felt rough, it indicates adulteration with sand. If hand is inserted in a bag or heap of cement, it must feel cool and not warm. If a small quantity of cement is thrown in a bucket of water, it must sink and should not float on the surface. A thin paste of cement with water must feel sticky between the fingers. If the cement contains too much of pounded clay and it as an adulterant, the paste can give an earthy smell.
- Presence of lumps: The cement must be free from any hard lumps. The lumps are formed by the absorption of moisture from the atmosphere. Any bag of cement containing such lumps must be rejected. This Field tests for cement are carried out to check lumps.
- Strength: This Field tests for cement is to check strength of cement and will be roughly ascertained in the following ways:
    - The briquettes with a lean or weak mortar are made. The size of briquette could be about 75mm x 25 mm x 12 mm. The proportion of cement and sand could be 1:6. The briquettes are immersed in water for a period of 3 days. If cement is of sound quality, such briquettes cannot be broken easily and it will be difficult to convert them into powder form.
    - A block of cement 25 mm x 25 mm and 200 mm long is prepared and it is immersed for 7 days in water. It is then placed on supports 150 mm apart and it is loaded with a weight of 340 N. The block should not indicate signs of failure.
    - A thick paste of cement with water is made on a piece of thick glass and it is kept under water for 24 hours. It must set and not crack.

Hydration

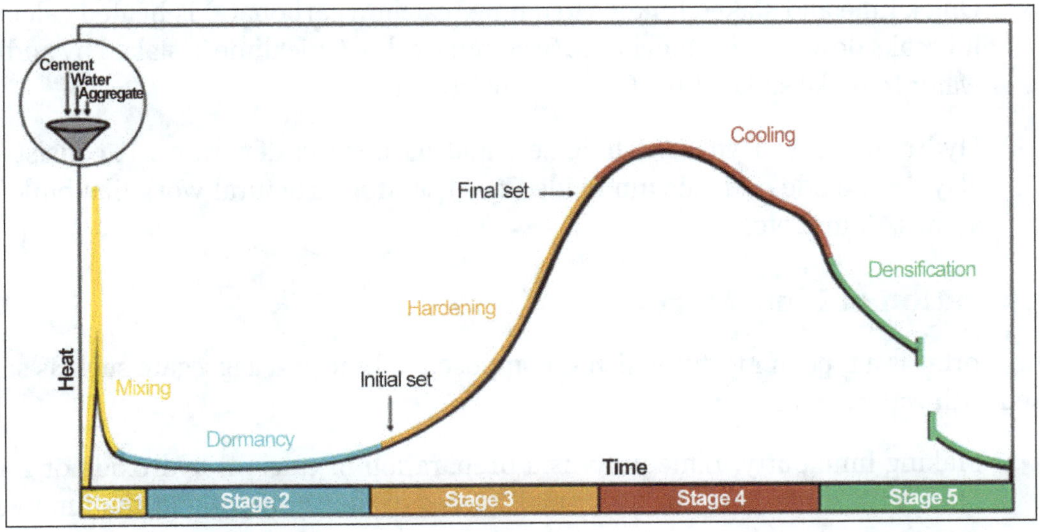

The chemical reaction between active components of cement and water is known as hydration of cement. It is an exothermic reaction. The heat generated in this process is called heat of hydration.

Compressive Strength

The compressive strength is the ability of a material or a structure to withstand loads that tends to reduce its size. The compressive strength can be measured by plotting the applied force against deformation in a compressor testing machine. Some materials may fracture at their compressive strength limit others deform irreversibly. Therefore a given amount of deformation may be considered as the limit for the compressive load. Compressive strength is a key value for design of structures.

Tensile Strength

Tensile strength is the maximum stress in which a material can withstand its load while being stretched or pulled before failing or breaking. Tensile strength is not as same as the compressive strength and the values can be quite different.

1.1.4 Lime

Lime is made from Calcium Carbonate ($CaCO_3$) which is found in chalk, limestone and sea shells.

Types of Lime

- Fat lime: Fat lime has high calcium content (95%). It sets hardens by absorbing CO_2 from air. Fat lime is used for white washing plastered walls and is added with sand & lime mortar for brick work.
- Quick lime and slaked lime: When lime (calcium carbonate) is heated calcined it breaks down to give quicklime (calcium oxide). Quicklime is slaked by adding water to make slaked lime (calcium hydroxide).
- Hydraulic lime: Hydraulic lime sets and hardens under water. It consists of clay, iron oxide and calcium oxide. It is used for structural work like building arches, domes etc.

Preparation of Lime Mortar

Lime mortar is a type of mortar which is composed of lime and aggregate such as sand mixed with water.

- Making lime putty. Lime putty is a preparation of calcium hydroxide or lime. Start with some lime chalk or small pieces of limestone. Blend the lime with water to create hydrated lime. Heat the hydrated lime at a temperature of about 1650 degrees Fahrenheit until it turns into a thick paste. Once the consistency is reached put the lime putty aside to set. Setting of lime normally takes several months. The end product should resemble any other common putty in its texture and hardness.

- Obtain a quantity of sand or other aggregate. Aggregate is the reagent which is added to the lime putty to make the actual mortar. Sand is the common but other materials like gravel can be used. Sometimes hair is added to the mix to help bind it together. Mix up the mortar according to the formula of 1 part lime putty to 3 parts aggregate. The brick and volcanic materials which is in powder can be added to the mixture to try to create something similar to the Portland cement. Portland cement is hydraulic which means it can set under water. Common lime mortar is non-hydraulic in nature.
- Laying the mortar by spreading it on top of the masonry and in between blocks of masonry. This can be bricks, stones or concrete blocks. Mortar takes time for drying. As it dries the calcium hydroxide which reacts with the air to produce the calcium carbonate. Calcium carbonate is actually the same as limestone. Limestone is the substance from which the lime putty was originally made. The material created is a hard bond or joint between the pieces of masonry.
- To obtain a faster drying mortar use hydrated lime in our mixture. This means heating the calcium hydroxide together with whatever impurities present in the original lime chalk. Usually this means bits of clay. Hydraulic lime reacts with water to form crystal. These crystals create a partial hardening effect that will be speed for the overall setting of the lime mortar.

1.1.5 Timber

The different timbers due to fungal action are:

- Dry Rot: Some fungi eat wood and cause it to crumble down as powder. This is known as dry rot. Fungi causing dry rot grow in dark, damp, sappy places. Seasoning prevents dry rot.
- Wet rot: When the timber's moisture content is above 30%, it is affected by fungi. This is known as wet rot.
- Heart rot: Fungi enter the tree through wounds in the hark and decay the heartwood.

Drying of wood is called seasoning. Seasoning wood is important because unseasoned wood shrinks during usage and may decay easily. Seasoning increases the strength, durability of timber.

Defects of Timber

A defect is an irregularity or abnormality occurring in or on wood which is responsible for its decrease in strength and durability.

Classification of Defects

1. Natural Forces

- Knots: These are sections of branches which are broken or cut off from the tree. The portion from which the branch is removed receives very less nourishment from the stem fora long time and it results in formation of dark hard rings which are known as knots. As continuity of wood fibers is broken by knots, they form a source of weakness in timber. Depending on the diameter knots are classified into:
    - Nail knots (less than 6 mm diameter).
    - Small knots (6 to 20 mm).
    - Medium knots (20 to 40 mm).
    - Large knots (more than 40 mm).
- Shakes: These are longitudinal separations in wood between the annual rings. These are cracks which partly or completely separate fibres of wood. The separations make the wood undesirable when appearance is important. Types of shakes are:
    - Cup shake.
    - Heart shake.
    - Ring shake.
    - Star shakes.
- Twisted grain or fibre: This defect is caused in the tree due to wind action.
- Upset or rupture: The upsets are due to an injury caused during the growth of tree due to strong winds or improper felling.
- Rind gall: Rind means bark and gall indicates abnormal growth. Hence peculiar curved swelling found on the body of tree is known as rind gall. They develop at points from where branches are improperly cut off or removed. The timber in this part is very weak and not durable.

2. Fungi and Insects

The important defects due to fungal actions are:

- Dry Rot: Certain species of fungi feed on woods, causing the wood to crumble down into powdery form. This is known as dry rot. The favorable conditions for growth of fungus causing dry rot are absence of sunlight, some dampness and presence of sap. Dry rot may be prevented by using well-seasoned sap free timber.

- Wet rot: When the moisture content of timber is more than 30%, it is affected by fungi. This is known as wet rot.
- Heart rot: Fungi enter the tree through wounds in the bark and decay the heart-wood. Defects due to insects are caused by beetles, marine borers and termites like white ants. They drill the wood forming holes into it and convert wood into powdered form.

1.2 Concrete

It is a composite material composed mainly of water, aggregate and cement. Often additives and reinforcements are included in the mixture to achieve the desired physical properties of the finished material. When these ingredients are mixed together they form a fluid mass that is easily molded into shape. Over time the cement forms a hard matrix which binds the rest of the ingredients together into a durable stone-like material with many uses.

Ingredients

The concrete consisting of cement, sand and coarse aggregates mixed in a suitable proportions in addition to water is called cement concrete. In this type of concrete cement is used as a binding material, sand as fine aggregates and gravel, crushed stones as coarse aggregates.

Admixtures are added before or during mixing of concrete in order to achieve certain goals which are explained below.

The two basic types of admixtures available are:

1. Mineral Admixture

They are usually added to concrete in larger amounts for the following reasons:

- To enhance the workability of fresh concrete.
- To improve resistance of concrete to thermal cracking, alkali- aggregate expansion and sulfate attack.
- To enable a reduction in cement content.

Examples of mineral admixture are fly ash, silica fume [SF] and ground granulated blast furnace slag.

2. Chemical Admixture

Chemical admixtures are added to concrete in very small amounts mainly for:

- Entrainment of air.
- Reduction of water or cement content.
- Plasticization of fresh concrete mixtures.
- Control of setting time.

Types of Chemical Admixtures

The different types of chemical admixtures are as follows:

1. Accelerating Admixture:

Accelerating admixtures (also known as accelerators) are added to concrete to reduce the setting time of concrete and to accelerate early strength. Accelerating admixtures are especially useful for modifying the properties of concrete in cold weather.

Examples: Calcium nitrate ($Ca(NO_3)_2$) and sodium nitrate ($NaNO_3$).

2. Retarding Admixtures:

Retarding admixtures are often used in hot weather conditions to delay the setting time of concrete. Examples: Calcium Sulphate ($CaSO_4$), sugar, starch.

3. Water reducing Admixture:

This type of admixture is added to reduce the water content by around 5% to 12% in a mixture while maintaining certain amount of consistency.

4. Plasticizers:

These are organic or a combination of organic and inorganic substances which allow water reduction for a given workability.

Examples: Hydro carbolic acid and lignosulfonate.

5. Super Plasticizers:

They are high range water reducers and reduce the water amount in cement by around 12% to 30%.

Examples: Melamine formaldehyde, naphthalene, sulphonate formaldehyde.

6. Air Entraining Admixtures:

Air entraining admixtures must be used whenever concrete is exposed to freezing and thawing. Air entraining agents entrain or trap in microscopic air bubbles in the concrete. When the hardened concrete freezes the frozen water inside the concrete expands into these air bubbles instead of damaging the concrete.

The air entrainment may be done by surface active agents (E.g. Natural wood resins, Animal or vegetable fats, oils such as olive oil) chemicals (Ex: zinc or aluminium powder) and cement dispersing agents (Ex: Calcium ligno sulphonate).

7. Bonding Admixtures:

They are added by 5% to 20% to concrete for increasing the bond strength between old and new concrete.

Ex: rubber, polyvinyl chloride, polyvinyl acetate, acrylic polymers.

The above structural adhesives can also be classified on the basis of the materials they are used.

Adhesives

Retarders

Retarders are an admixture which slows the process of hydration. Concrete remains plastic and workable because of retarders. If concreting is done in hot weather, retarders overcome the accelerating effect of high temperature. Retarders are used in consolidating a large number of pours without the formation of cold joints and in grouting oil wells. Calcium sulphate is the commonly used retarder.

Admixtures increase the compressive strength by 10 to 20 %. Retarding plasticizers are available in the market. These are an important type of admixtures often used in the ready mixed concrete industry, for retaining the slump loss, during high temperature, long transportation and to avoid construction cold joints.

Accelerators

It is useful type of admixture which are added to get the early strength. Such situations may occur under the following conditions:

- When early removal of form work is needed.
- When reduction of period of curing is needed.
- When to put the structure early to use.

- When accelerating the setting time in cold weather.

Commonly used accelerator was calcium chloride. Soluble carbonates, silicates, fluosilicate and some of the organic compounds are used. Fluosilicates and organic compound like trietheuolamine are comparatively expensive. Some of the accelerators available now can make the cement set into stone hard in a matter of five minutes. The availability of such accelerators for underwater concreting has become easy. Waterfront structures which need repair in short time may be done using accelerators. These materials could be used in cold environments up to 60°C.

Manufacturing Process

A good quality concrete is essentially a homogeneous mixture of cement, coarse and fine aggregates and water which consolidates into a hard mass due to chemical action between the cement and water. Each of the four constituents has a specific function. The coarser aggregate acts as a filler. The fine aggregate fills up the voids between the paste and the coarse aggregate. The cement in conjunction with water acts as a binder.

The mobility of the mixture is aided by the cement paste fines and nowadays increasingly by the use of admixtures. Most of the properties of the hardened concrete depend on the care exercised at every stage of the manufacture of concrete. A rational proportioning of the ingredients of concrete is the essence of the mix design. However it may not guarantee of having achieved the objective of the quality concrete work. The aim of quality control is to ensure the production of concrete of uniform strength from batch to batch. This requires some rules to be followed in the various stages of concrete production and are discussed as follows.

The stages of concrete production are:

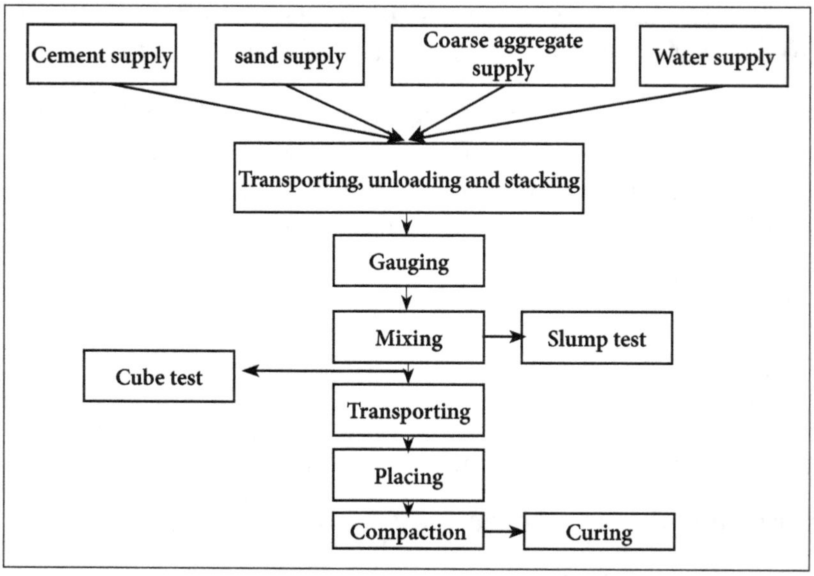

Manufacturing process of concrete.

1. Batching of Materials

For good quality concrete a proper and accurate quantity of all the ingredients should be used. The aggregates, cement and water should be measured with an accuracy of ± 3 per cent of batch quantity and the admixtures by 5 per cent of the batch quantity. There are two prevalent methods of batching materials the volume batching and the weigh batching. The factors affecting the choice of batching method are the size of job, required production rate and required standards of batching performance. For most important works weigh batching is recommended.

- Volume Batching.
- Weigh Batching.

2. Mixing

Mixing concrete is simply defined as the complete blending of the materials which are required for the production of a homogeneous concrete.

- Hand Mixing.
- Machine Mixing:
    - Tilting Mixers.
    - Non -tilting Mixer.
    - Reversing Drum Mixer.
    - Pan-type or Stirring Mixer.
    - Transit Mixer.

Charging the Mixer and Mixing Time: The order of feeding the ingredients into the mixer is as follows:

About 25 per cent of water required for mixing is first introduced into the mixer drum to prevent any sticking of cement on the blades and bottom of the drum. Then the ingredients are discharged through the skip.

In the skip the sequence of loading should be to add first half the coarse aggregate then half the fine aggregate and over this total cement and then the balance aggregates. After discharging the ingredients into the drum the balance water is introduced. The mixing time is counted from the instant complete water is fed into the mixer. The speed of the mixers is generally 15 to 20 rpm. For proper mixing the number of revolutions per minute required by the drum are 25 to 30. Time of mixing also depends on capacity of mixer and is given in table:

Time of Mixing		
Capacity of Mixer	Minimum mixing time (minutes)	
(cum)	Natural aggregates	Manufactured aggregates

³3	2	2.5
2	1.5	2
£1	1.25	1.5

A poor quality of concrete is obtained if the mixing time is reduced. On the other hand if the mixing time is increased it is uneconomical. However it is found that if the mixing time is increased to 2 minutes the compressive strength of concrete produced is enhanced and beyond this time the improvement in compressive strength is insignificant. A prolonged mixing may cause segregation. Also due to longer mixing periods the water may get absorbed by the aggregates or evaporate resulting in loss of workability and strength.

3. Transporting

Concrete should be transported to the place of deposition at the earliest without the loss of homogeneity obtained at the time of mixing. A maximum of 2 hours from the time of mixing is permitted if trucks with agitator and 1 hour if trucks without agitators are used for transporting concrete. Also it should be ensured that segregation does not take place during transportation and placement. The methods adopted for transporting concrete depend upon the size and importance of the job, the distance of the deposition place from the mixing place and the nature of the terrain. Some of the methods of transporting concrete are as below:

- Mortar Pan.
- Wheel Barrow.
- Chutes.
- Dumper.
- Bucket and Rope way.
- Belt conveyor.
- Skip and Hoist.
- Pumping.

4. Placing

To achieve quality concrete it should be placed with utmost care securing the homogeneity achieved during mixing and the avoidance of segregation in transporting. Research has shown that a delayed placing of concrete results in a gain in ultimate compressive strength provided the concrete can be adequately compacted. For dry mixes in hot weather delay of half to one hour is allowed whereas for wet mixes in cold weather it may be several hours.

5. Curing

- Pouring water.
- Covering with damp gunny bags and sprinkle water regularly over it.
- Steam curing: Curing can be also accomplished by artificial heat while the concrete is maintained in moist condition. This method of curing is also known as accelerated curing since an increased rate of strength development can be achieved.
- Completely immersing it in water tanks.

Water is needed to chemically react with the cement (hydration) and provide workability to the concrete. The amount of water in the mix compared with the amount of cement is called the water/ cement ratio. Lower the w/c ratio is maintaining the workability of concrete.

Workability: Factors Affecting Workability

It is the property which determines the ease and homogeneity with which concrete can be mixed, placed, compacted and finished. A workable concrete will not have any segregation or bleeding. Segregation causes large voids and hence concrete becomes less durable. Bleeding results in several small pores on the surface due to excess water coming up. Bleeding also makes concrete less durable. The degree of workability of concrete is classified from very low to very high with the corresponding value of slump in mm.

Water Content or Water Cement Ratio

More the water cement ratio more will be workability of concrete. Since by simply adding water the inter particle lubrication is increased. High water content results in a higher fluidity and greater workability but decreases the strength of concrete. Because with increasing w/c ratio the strength decreases as more water will result in higher concrete porosity. So, the lower the w/c, the lower is the void volume/solid volume and the stronger the hardened cement paste.

Increased water content also results in bleeding, hence, increased water content may also mean that cement slurry can escape through the joints of the formwork.

Influence of Aggregate Mix Proportions

Increasing the proportion of aggregates relative to the cement can decrease the workability of the concrete. Also, any additional fines will require more cement within the mix. An "over sanded" mix will be permeable and less economical. A concrete deficient of fines will be difficult to finish and prone to segregation.

Aggregate Properties

The ratio of coarse/fine aggregate is not the only factor affecting workability. The gradation and particle size of sands are important. Shape and texture of aggregate will also affect workability. Spherical shaped particles will not have the interaction problems associated with more angular particles. Also, spherical shapes have a low surface/volume ratio, therefore, less cement will be required to coat each particle and more will be available to contribute to the workability of the concrete. Aggregate that is porous will absorb more water leaving less to provide workability. This is important to distinguish between total water content, that includes absorbed water, and free water which is available for improving workability.

Loss of Workability

Workability will decrease with time due to several factors and continued slow hydration of C3S and C_3A during dormant period, loss of water through evaporation and absorption, increased particle interaction due to the formation of hydration products on the particle surface. Loss of workability is measured as "slump loss" with time.

Admixtures

In usual, air-entraining, water-reducing, and set-retarding admixtures will all improve workability. However, some chemical admixtures can react differently with cements and aggregates and should result in reduced workability.

Properties of Hardened Concrete

1. Strength

Strength is achieved by standard cubes after 28 days of casting and it is known as the characteristic strength of mix and is used for design purposes. Actually the strength of concrete goes on increasing even after 28 days of curing. Hence on the member which is not likely to get the full load within 28 days (e.g. foundation concrete) the characteristic strength may be increased by the age factor which is given in the table:

Minimum age of member when design load is expected (in months)	Age factor
1	1.0
3	1.10
6	1.15
12	1.20

The tensile strength may be calculated from the formula.

$$F_{cr} = 0.7\sqrt{fck}.$$

Where fck is the characteristic compressive stress.

The modulus of elasticity may be estimated from the formula.

$$E = 5700\sqrt{fck}\ N/mm2.$$

2. Dimensional Changes

Concrete gets shrinkage with its age. The permanent size depends upon the constituents of concrete, size of the member and the environmental conditions. The total shrinkage may be taken as 0.0003 approximately. The permanent dimensional changes that take place due to loading is known as creep. Its value depends upon the stresses in concrete, the age of concrete at loading and the duration of loading. The ultimate creep strain may be estimated from the values of creep coefficient which is listed in table:

Table: Creep coefficient based on loading age

Age of loading	Creep coefficient
7 days	2.2
28 days	1.6
1 year	1.1

The size of concrete may change due to thermal expansion. The coefficient of thermal expansion depends on the nature of cement, the aggregate, the cement content, the relative humidity and the size of sections.

3. Durability

Durability refers to the resistance to the forces of environment such as weathering, chemical attack and fire. Freezing and thawing results in adverse stresses which develops due to the presence of water in the pores of the concrete. Generally dense and strong concretes have better durability in extreme weather conditions. The cube crushing strength is only not a reliable guide to the quality and durability of concrete. It should have an adequate cement content and a low water-cement ratio.

4. Impermeability

Impermeability is the resistance of the concrete to the flow of water through the pores. Excess water during the manufacturing of concrete leaves a large number of continuous pores leading to permeability. Since permeability reduces the durability of concrete it should be kept very low by using low water-cement ratio, dense and well-graded aggregates, proper compaction and continuous curing at low temperature conditions. The cement content used should be sufficient enough to provide the adequate workability with a low water cement ratio and with the available compaction method.

Types of Mixes

1. Nominal Mixes

In the past the specifications for concrete prescribed the proportions of cement, fine and coarse aggregates. These mixes of fixed cement-aggregate ratio which ensures adequate strength are termed nominal mixes. These offer simplicity and under normal circumstances have a margin of strength above that specified. However due to the variability of mix ingredients the nominal concrete for a given workability varies widely in strength.

2. Standard Mixes

The nominal mixes of fixed cement-aggregate ratio (by volume) vary widely in strength and may result in under- or over-rich mixes. For this reason the minimum compressive strength has been included in many specifications. These mixes are termed standard mixes.

IS 456-2000 has designated the concrete mixes into a number of grades as M_{10}, M_{15}, M_{20}, M_{25}, M_{30}, M_{35} and M_{40}. In this designation the letter M refers to the mix and the number to the specified 28 day cube strength of mix in N/mm^2. The mixes of grades M_{10}, M_{15}, M_{20} and M_{25} correspond approximately to the mix proportions (1:3:6), (1:2:4), (1:1.5:3) and (1:1:2) respectively.

3. Designed Mixes

In these mixes the performance of the concrete is specified by the designer but the mix proportions are determined by the producer of concrete except that the minimum cement content can be laid down. This is most rational approach to the selection of mix proportions with specific materials in mind possessing more or less unique characteristics. The approach results in the production of concrete with the appropriate properties most economically. However the designed mix does not serve as a guide since this does not guarantee the correct mix proportions for the prescribed performance.

For the concrete with undemanding performance nominal or standard mixes (prescribed in the codes by quantities of dry ingredients per cubic meter and by slump) may be used only for very small jobs when the 28-day strength of concrete does not exceed 30 N/mm^2. No control testing is necessary reliance being placed on the masses of the ingredients.

Relation between Compression and Tensile Strength

1. Compressive Strength

The main measure of the structural quality of concrete is its compressive strength. This property of concrete is commonly considered in structural design. Depending on the mix and time and quality of the curing, compressive strength of concrete can be obtained up to 14,000 psi or more. Commercial production of concrete with ordinary aggregate is usually in the 3,000 to 12,000 psi range with the most common ranges for cast-in-place buildings from 3,000 to 6,000 psi. On the other hand, precast and prestressed applications often expect strengths of 4,000-8,000 psi.

2. Tension Strength

Though concrete is not normally designed to resist direct tension, the knowledge of tensile strength is used to estimate the load under which cracking will develop. This is due to its influence on the formation of cracks and its propagation to the tension side of reinforced concrete flexural member. Shear, torsion and other actions also produce tensile stresses to the particular section of concrete member. In most cases member behavior changes upon cracking. So tension strength of concrete is also considered in proportioning concrete member. This strength is of interest in designing of highway and airfield slabs as shear strength and resistance to cracking are very important to sustain such loading. The tensile strength of concrete is relatively low, about 10 to 15% of the compressive, occasionally 20%.

Special Concretes

Shotcrete

The concrete in which mortar or concrete is pneumatically projected at high velocity on the backup surface is known as shotcrete or gunite. It is also called spray concrete as the force of jet impacting on the surface compacts it so as to make itself supporting. This type of concrete has no special properties as compared to normal concrete placed under similar conditions.

Advantages:

- Shotcrete concrete layers are very strong.
- These types of concrete do not need construction or expansion joints.
- It can be evenly applied on uneven surfaces and can be applied from a distance.

Disadvantages:

- It has high cost.
- Its lining is less durable than ordinary concrete lining of the same thickness.

Light Weight Aggregate Concrete

Lightweight concretes will either be lightweight aggregate concrete, foamed concrete or autoclaved aerated concrete (AAC). Such lightweight concrete blocks are often used in house construction. Lightweight aggregates originate from either:

- Natural materials, like volcanic pumice.
- The thermal treatment of natural raw materials like clay, slate or shale.
- Manufacture from industrial by-products such as fly ash.
- Processing of industrial by-products like FBA or slag.
- The required properties of the lightweight concrete will have a bearing on the best type of lightweight aggregate to use. If little structural requirement, but high thermal insulation properties, are needed then a light, weak aggregate will be used. This will result in relatively low strength concrete.

Foamed Concrete

Foamed concrete is a highly workable, low-density material which can incorporate up to 50 per cent entrained air. It is generally self-levelling, self-compacting and may be pumped. Foamed concrete is ideal for filling redundant voids such as disused fuel tanks, sewer systems, pipelines, and culverts - particularly where access is difficult. It is a recognized medium for the reinstatement of temporary road trenches. Good thermal insulation properties make foamed concrete also suitable for sub-screeds and filling under-floor voids.

Autoclaved Aerated Concrete (AAC)

AAC was first commercially produced in 1923 in Sweden. Since then, AAC construction systems such as masonry units, reinforced floor/roof and wall panels and lintels have been used on all continents and every climatic condition. AAC can also be seen by hand, sculpted and penetrated by nails, screws and fixings.

Cellular Concrete

Cellular concrete is a lightweight cement-based material, containing many gas bubbles evenly distributed in the volume, produced by blending and maturing of a mixture of cement, filler, water, agent generating cells.

By the method of generating the air or gas cells there exist foam concrete and gas concrete. Gas concrete is presented mainly in the form of pre-cast blocks, the majority of its applications however demand it to be cast-in-place.

Cellular concrete is produced with various densities from 250 kg/m^3 to 1600 kg/m^3. CFC is harder in comparison with traditional insulation materials, especially when chemical and fire resistance is needed.

High Density Concrete

High density or heavyweight concrete is concrete with a density greater than 2600kg/ m^3. Its primary use is in radiation shielding, either in nuclear power plants or in radiation therapy units. It can also be used as ballast in offshore locations such as pipelines.

High density concrete can be made from natural heavyweight aggregates such as barites or magnetite which typically give densities of 3500kg/m^3 and 3900kg/m^3respectively. They can also be made using iron or lead as a replacement for a portion of the aggregate. These give even greater densities of 5900kg/m^3 for iron or 8900kg/m^3 for lead.

Fibre Reinforced Concrete

- Fibre reinforced concrete (FRC) may be defined as a composite materials made with Portland cement, aggregate, and incorporating discrete discontinuous fibres.
- The role of randomly distributes discontinuous fibres is to bridge across the cracks that develop provides some post- cracking "ductility".
- The real contribution of the fibres is to increase the toughness of the concrete under any kind of loading.
- The fibre reinforcement may be used in the form of three – dimensionally randomly distributed fibres throughout the structural member when the added advantages of the fibre to shear resistance and crack control can be further utilized.

Factors Affecting Properties of F.R.C

1. Volume of Fibre:

- Low volume fraction (less than 1%).
- Used in slab and pavement that have large exposed surface leading to high shrinkage cracking.
- Moderate volume fraction (between 1 and 2 percent).
- Used in Construction method such as Shortcrete & in Structures which requires improved capacity against delamination, spelling & fatigue?
- High volume fraction (greater than 2%).
- Used in making high performance fibre reinforced composites (HPFRC).

2. Aspect Ratio of Fibre:

- It is defined as ratio of length of fibre to it's diameter (Lid).

- Increase in the aspect ratio up to 75, there is increase in relative strength and toughness. Beyond 75 of aspect ratio there is decrease in aspect ratio and toughness.

3. Orientation of Fibres:

- Aligned in the direction of load.
- Aligned in the direction perpendicular to load.
- Randomly distribution of fibres.
- It is observed that fibres aligned parallel to applied load offered more tensile strength and toughness than randomly distributed or perpendicular fibre.

4. Relative Fibre Matrix:

- Modulus of elasticity of matrix must be less than of fibres for efficient stress transfer. Low modulus of fibres imparts more energy absorption while high modulus fibres imports strength and stiffness.
- Low modulus fibres e.g. Nylons and Polypropylene fibres.
- High modulus fibres e.g. Steel, Glass, and Carbon fibres.

Tests

The compression strength test is used for specifying the grade of concrete in design and for quality control of field concrete.

The test specimens recommended are 150 × 150 × 150 mm cubes or cylinders of 150 mm diameter and 300 mm height. The mixed concrete is filled into the moulds in layers of 50 mm to achieve full compaction. Each layer of mix is tamped using a bar or vibrator. The test specimens are stored at a temperature of 27± 3°C and at 90 per cent humidity for about 1 day. After this period the specimens are removed from the moulds and placed in water for 7 and 28 days.

At least three specimens preferably from different batches are tested at each selected age. The specimens should be tested immediately after taking them from water with surface water wiped off.

Now the cube is placed in the machine and load is applied. The rate of loading should be approximately 14 N/mm^2 per minute till the failure of the cube.

$$\text{Flow}(\%) = \frac{\text{Spread dia.(cm)} - 25}{25} \times 100$$

The average of the three values is taken as the compressive strength of concrete of the batch provided the individual variation is not more than ± 15 per cent of the average.

1. Compaction Factor Test

- The sample of concrete is placed in the upper hopper up to the brim.
- The trap-door is opened so that the concrete falls into the lower hopper.
- The trap-door of the lower hopper is opened and the concrete is allowed to fall into the cylinder.
- The excess concrete remaining above the top level of the cylinder is then cut off with the help of plane blades.
- The concrete in the cylinder is weighed. This is known as weight of partially compacted concrete.

The cylinder is filled with a fresh sample of concrete and vibrated to obtain full compaction. The concrete in the cylinder is weighed again. This weight is known as the weight of fully compacted concrete.

Compacting factor = (Weight of partially compacted concrete)/ (Weight of fully compacted concrete)

Higher is the compaction factor greater the workability.

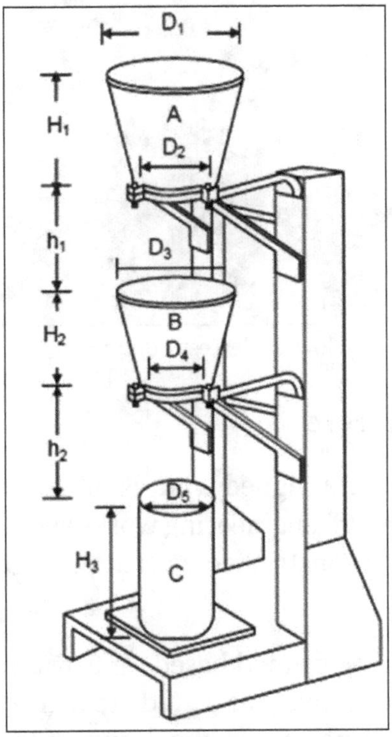

Compaction Factor Test Specimen.

2. Flow Table Test

- This test is used for high workability concrete with a slump of more than 150mm.
- The mould is firmly held on the centre of the table and filled with concrete in two layers each approximately one-half the volume of the mould and tamped 25 times with a rod in a uniform manner.
- The mould is then removed from the concrete by a steady upward pull.
- The table is then raised and dropped from a height of 12.5 mm, 15 times in about 15 seconds.
- The diameter of the spread concrete is measured.

The flow of the concrete is the percentage increase in diameter of spread concrete over the base diameter of the moulded concrete calculated from the following formula:

$$\text{Flow}(\%) = \frac{\text{Spread dia.(cm)} - 25}{25} \times 100$$

Flow Table Test Equipment.

Quality Control of Concrete

Quality control implies that the assigned work is done according to the specifications agreed in the contract. Major civil engineering works such as multi -storeyed buildings, dams, harbors, etc. have to be constructed with at most care as they have to be in use for decades.

Specifications of work should be framed based on Government or standard procedure such that they have to serve effectively as a guide to complete the work with high quality. The specifications are as important as the design of the project. In order to make a quality concrete construction at a site the field work has to be organized with the three divisions with mutual coordination, viz., engineering division, manufacturing division and placing division.

Engineering division looks after all forms, reinforcement details and installations of all embedded parts. Manufacturing division takes care of the control of materials, batching and mixing of concrete. Placing division takes care of placing during and other subsequent works. Whole aim is to produce a quality concrete of high order and with high economics.

Requirements for High Quality Concrete

- Air bubbles should be completely removed from the concrete.
- Compaction of concrete should be such that a minimum void is present.
- Adequate curing for 28 days has to be effected.

Advantages of Carefully Constructed Concrete

- Failure possibility is minimized.
- Lower cost of construction with long life.
- Low maintenance cost.

Statistical Methods

In all statistical observations, it is necessary to get an idea of the central tendency of the variable being studied. The following are some of the properties used to represent this tendency:

- Median: It is the mid value or that value which has as many values greater than itself as there are less than itself.
- Mode: It is the most commonly occurring value in the set of values being considered.
- Mean value or the arithmetic mean: It is the arithmetic average of all the values being considered.

The spread of the curve is practically between -3s and -3s, but theoretically it is between -cc and +Q.. From the per cent of areas represented for various values of z given in Figures. We can see that the value of z corresponding to 5 per cent failure is given by z = (fck - fmz)/ s = 1.64 and the failure corresponding to z = 2.0 is 2.3%. Similarly, the failure corresponding to z = 2.33 is about 1%, and that corresponding to = 3 is about 0.1% only.

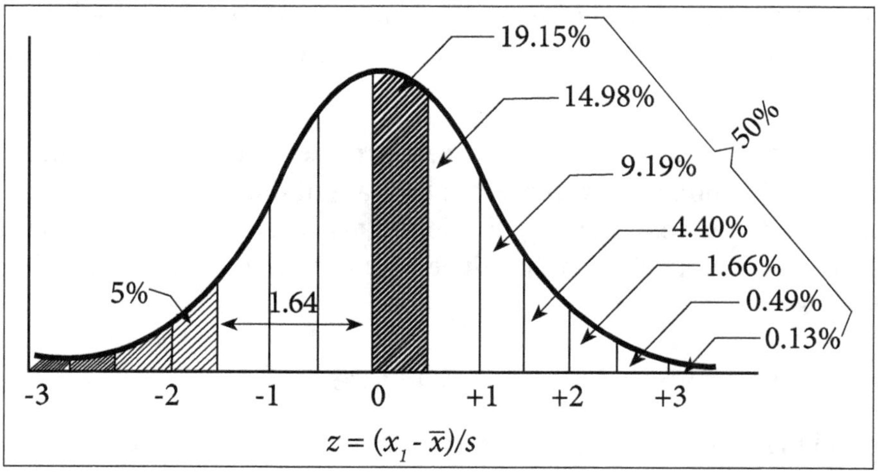

Areas under normal probability rune.

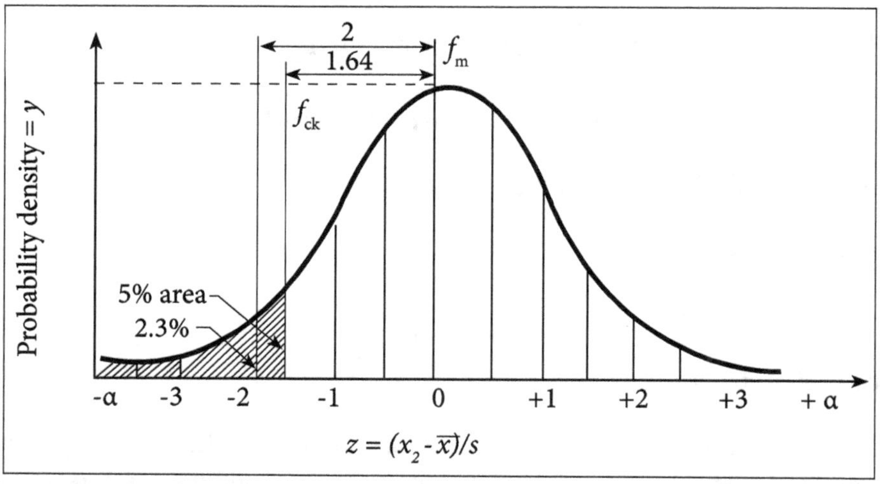

Illustration or the principle used fur arriving at target strength for a specified characteristic strength.

Acceptance Criteria

1. The concrete shall be deemed to comply with the strength requirements if:

- Every sample has test strength not less than the characteristic value.
- The strength of one or more samples though less than the characteristic value, is in each case not less than the greater of:
  - The characteristic strength minus 1.35 times the standard deviation.
  - 0.80 times the characteristic strength and the average strength of all the samples is not less than the characteristic strength plus.

$$p = 100\left(\frac{d}{D}\right)^{1/2}$$

2. The concrete shall be deemed not to comply with the strength requirements if:

- The strength of any sample is less than the greater of:
    - The characteristic strength mix is 1.35 times the standard deviation.
    - 0.80 times the characteristic strength.
- The average strength of all samples is less than the characteristic strength plus.

$$\left[1.65-\frac{3}{\sqrt{\text{No.of samples}}}\right] \text{times the standard deviation.}$$

3. Concrete which does not meet the strength requirements as specified in I, but has strength greater than that required by II may, at the discretion of the designer, be accepted as being structurally adequate without further testing.

4. Concrete of each grade shall be assessed separately.

5. Concrete shall be assessed daily for compliance.

6. Concrete is liable to be rejected if it is porous or honey-combed; its placing has been interrupted without providing a proper construction joint the reinforcement has been displaced beyond the tolerances specified or construction tolerances have not been met. However, the hardened concrete may be accepted after carrying out suitable remedial measures to the satisfaction.

7. Where the value of the average strength of the tests is less than shall be rejected.

$$f\,ck+\left\{1.65-\frac{1.65}{\eta}\right\}\times s$$

Concepts Proportioning of Concrete Mixes by Various Methods

1. Arbitrary Method

The general expression for the proportions of cement, sand and coarse aggregate is 1: n: 2n by volume.

1: 1: 2 and 1: 1.2: 2.4 for very high strength.

1: 1.5: 3 and 1: 2: 4 for normal works.

1: 3: 6 and 1: 4: 8 for foundations and mass concrete works.

Recommended Mixes of Concrete: The concrete as per IS 456: 2000, the grades of concrete lower than M20 are not to be used in RCC work.

M10	1:3:6
M15	1:2:4
M20	1:1.5:3
M25	1:1:2

2. Fineness Modulus Method

The term fineness modulus is used to indicate an index number which is roughly proportional to the average size of the particle in the entire quantity of aggregates.

The fineness modulus is obtained by adding the percentage of weight of the material retained on the following sieve and divided by 100.

The coarser the aggregates, the higher the fineness modulus.

Sieve is adopted for:

- All aggregates: 80 mm, 40 mm, 20 mm, 10 mm, and Nos. 480, 240, 120, 60, 30 and 15.
- Coarse aggregates: mm, 40 mm, 20 mm, 10 mm, and No. 480.
- Fine aggregates: Nos. 480, 240, 120, 60, 30 and 15.

Proportion of the fine aggregate to the combined aggregate by weight:

$$R = \frac{P \quad P}{P \quad P} \times 100$$

Where, P = desired fineness modulus for a concrete mix of fine and coarse aggregates.

P_1 = fineness modulus of fine aggregate.

P_2 = fineness modulus of coarse aggregate.

3. Minimum Void Method

The quantity of sand used should be such that it completely fills the voids of coarse aggregate. Similarly, the quantity of cement used shown such that it fills the voids of sand, so that a dense mix the minimum voids is obtained.

In actual practice, the quantity of fine aggregate used in the mix is about 10% more than the voids in the coarse aggregate and the quantity of cement is kept as about 15% more than the voids in the fine aggregate.

4. Maximum Density Method

$$p = 100\left(\frac{d}{D}\right)^{1/2}$$

Where, D = maximum size of aggregate (i.e. coarse aggregate).

P = percentage of material finer than diameter d (by weight).

d = maximum size of fine aggregate.

A box is filled with varying proportions of fine and coarse aggregates. The proportion which gives heaviest weight is then adopted.

5. Water: Cement Ratio Method

According to the water – cement ratio law given by Abram as a result of many experiments, the strength of well compacted concrete with good workability is dependent only on the ratio.

The lower water content produces stiff paste having greater binding property and hence the lowering the water-cement ratio within certain limits results in the increased strength.

Similarly, the higher water content increases the workability, but lower the strength of concrete.

The optimum water-cement ratio for the concrete of required compressive strength is decided from graphs and expressions developed from various experiments.

Amount of water less than the optimum water decreases the strength and about 10% less may be insufficient to ensure complete setting of cement. An increase of 10% above the optimum may decrease the strength approximately by 15% while an increase in 50% may decrease the strength to one-half.

According to Abram's Law water-cement law, lesser the water-cement ratio in a workable mix greater will be the strength.

If water cement ratio is less than 0.4 to 0.5, complete hydration will not be secured.

Some practical values of water cement ratio for structure reinforced concrete:

- 0.45 for 1: 1: 2 concrete
- 0.5 for 1 : 1.5 : 3 concrete
- 0.5 to 0.6 for 1: 2 : 4 concrete.

Concrete vibrated by efficient mechanical vibrators require less water cement ratio, and hence have more strength.

Thumb Rules for deciding the quantity of water in concrete:

- Weight of water = 28% of the weight of cement + 4% of the weight of total aggregate.

- Weight of water = 30% of the weight of cement + 5% of the weight of total aggregate.

1.3 Elements of Building Construction

The functions of these components and the main requirement of them is presented as follows:

1. Foundation

Foundation is the most important part of the building. Building activity starts with digging the ground for foundation and then building it. It is the lower most part of the building. It transfers the load of the building to the ground. Its main functions and needs are:

- Distribute the load from the structure to soil evenly and safely.
- To anchor the building to the ground so that under lateral loads building cannot move.
- It provides level surface for the construction of super structure.

2. Plinth

The portion of the wall between the ground level and the ground floor level is known as plinth. It is usually of stone masonry. If the foundation is on piles a plinth beam is cast to support wall above floor level.

At the top of plinth a damp proof course is provided. It is typically 75 mm thick plain concrete course. The function of the plinth is to keep the ground floor above ground level free of dampness. Its height is not less than 450 mm. It is required that plinth level is at least 150 mm above the road level so that connections to underground drainage system will be made.

3. Walls and Columns

The function of walls and columns is to transfer the load of the structure vertically downwards to transfer it to foundation. Apart from this wall performs the following functions also:

- It encloses building area into different compartments and provides privacy.
- It gives safety from burglary and insects.
- It keeps the building warm in winter and cool in summer.

4. Sills, Lintels and Chajjas

A window frame should not be directly placed over masonry. It is placed over 50 mm to 75 mm thick plain concrete course provided over the masonry. This course is called sill. Lintels are the R.C.C. or stone beams provided over the door and window openings to transfer the load transversely so as to see that door or window frame is not stressed unduly.

The width of lintels is equal to the width of wall while thickness to be provided depends upon the opening size. Chajjas is the projection given outside the wall to protect doors and windows from the rain. They are typically made with R.C.C. In low cost houses stone slabs are provided as Chajjas.

The projection of Chajjas varies from 600 mm to 800 mm. Sometimes drops are also provided to Chajjas to improve aesthetic look and also to get additional protection from sun and rain.

5. Doors and Windows

The function of a door is to give access to various rooms in the building and to deny the access whenever necessary. Number of doors should be minimum possible. The size of the door should be of such dimension as will facilitate the movement of the largest object likely to use the door.

Windows are provided to get light and ventilation in the building. They are located at a height of 0.75 m to 0.9 m from the floor level. In hot and humid regions, the window area must be 15 to 20 percent of the floor area. Another thumb rule used to determine the size and the number of windows is for every 30 m^3 of inside volume there should be 1m^2 window opening.

6. Floors

These are the important component of a building. They give working/useful area for the occupants. The ground floor is prepared by filling brick bats, waste stones, gravel and well compacted with not less than 100 mm sand layer on its top.

A lean concrete of 1: 4: 8, 100 mm thick is laid. On this a damp proof courses are provided. Then floor finishing is done as per the requirement of the owner. Cheapest floor finish for a moderate house is with 20 to 25 mm rich mortar course finished with red oxide. The costliest floor finish is mosaic or marble finishing.

7. Roof

Roof is the top most portion of the building which provide top cover to the building. It must be leak proof. Sloping roof like tiled and A.C. sheet give leak proof cover easily. But they do not give provision for the construction of additional floor. Tiled roof give good thermal protection.

Flat roofs give provision for additional floors. Terrace adds to the comfort of occupants. Water tanks will be easily placed over the flat roofs.

8. Step, Stairs and Lifts

Steps provide convenient access from ground level to ground floor level. They are required at doors in the outer wall. 250 to 300 mm wide and 150 mm rise is ideal size for steps. In case the size of two consecutive steps be different.

Number of steps required depends upon the difference in the levels of the ground and the floor. Stairs give access from floor to floor. They must consists of steps of uniform sizes.

In all public buildings lifts are to be provided for the conveniences of old and disabled persons. In hostels G + 3 floors can be built without lifts but in residential flats maximum floors permitted without lifts is only G + 2. Lift is to be located near the entrance. Size of the lift is decided by the number of users in peak hours. Lifts are available with capacity 4 to 20 persons.

9. Finishing

Bottom portion of slab (ceiling), walls and top of floor need smooth finishing with plaster. Then they are provided with white wash, distemper or paints or tiles. The function of finishing work is:

- Give protective cover.
- Improve aesthetic view.
- Rectify defective workmanship.
- Finishing work for plinth consists in pointing while for floor it consists in polishing.

10. Building Services

Water supply, sanitation and drainage works, electric supply work and construction of cupboards and show cases constitute major building services.

For storing water from municipal supply or from tanker a sump is built in the house property near street. From the sump water is pumped to overhead tanks placed on or above roof level so as to get water all the 24 hours. Plumbing work is made so as to get water in kitchen, bathrooms, water closets, sinks and garden taps.

For draining rain water from roofs down take pipes of at least 100 mm diameters must be used. Proper slopes should be given to roof towards down take pipe. These pipes should be fixed at 10 to 15 mm below the roof surface so that rain water is directed to the down take pipe easily.

The sanitary fittings are to be connected to stone ware pipes with suitable traps and chambers. Stone ware pipes are then connected to underground drainage of municipal lines or to the septic tank.

Several carpentry works are required for building service. They are in the form of show-cases, cupboards, racks etc. Electric supply is essential part of building services. The building should be provided with sufficient points for supply of lights, fans and other electric gadgets.

Basic Requirements of a Building

The planning and construction of a building should be aimed at fulfilling the following requirements:

- Strength and stability.
- Dimensional stability.
- Resistance to dampness.
- Resistance to fire.
- Heat insulation.
- Sound insulation.
- Protection against termite attack.
- Durability.
- Security against burglary.
- Lighting and ventilation.
- Comforts and convenience.
- Economy.

1.4 Foundations Conventional Spread Footings and RCC Footings

Conventional Spread Footings

This kind of foundations is commonly used for walls and masonry columns. These foundations are built after opening the trenches to required depth. Such footings are economical up to a maximum depth of 3 m. As these foundations are suitable depth they are grouped under shallow foundations.

Figure illustrates a conventional spread footing for a wall and Figure shows it for a masonry column:

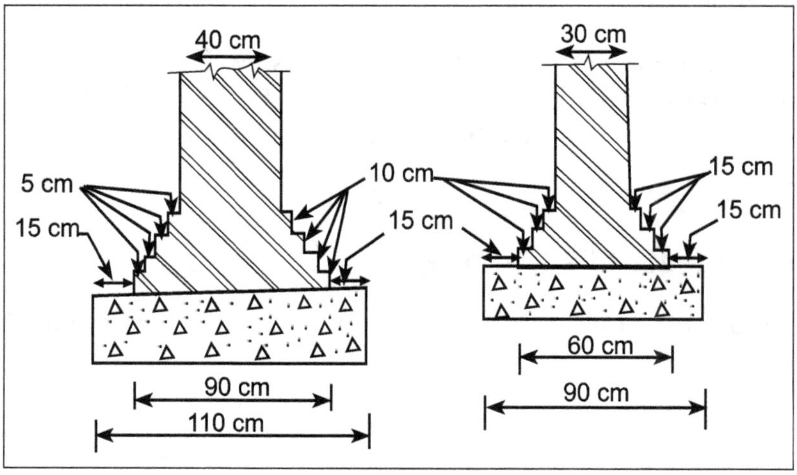

Wall footing.

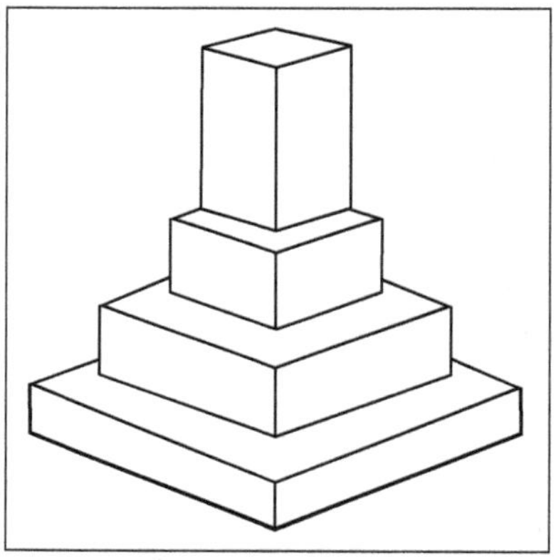

Foundation for masonry pier.

Before building these footing trenches are opened to required depth and the soil is rammed well. Then a plain concrete of mix 1: 4: 8 is provided. Its thickness varies from 150 to 200 mm. Over this bed, stone masonry footing is built. It is built in courses each course projecting 50 to 75 mm from the top course and height of each course being 150 to 200 mm. In case of wall footing the projections are only one direction while in case of columns they are in both directions. The projection of bed concrete from the lowest course of foundation masonry is typically 150 mm.

R.C.C. Footings

There are mainly two types of R.C.C. footings:

1. One Way Reinforced Footing: These footings are used for the walls. In these footings main reinforcements are in the transverse direction of wall. In longitudinal directions there will be only nominal reinforcement.

2. Two Way Reinforced Footings: For columns two way reinforced footings are provided. The following kinds of the footings are common:

- Isolated Column Footings: If separate footings are provided for each column it is known as isolated column footing. Figure shows a typical isolated column footing. The size of footing is based on the area required to distribute the load of the columns safely over the soil. These footings are provided over a 100 to 150 mm bed concrete. Required reinforcements and thickness of footing are found by the design engineers. Thickness is uniform or varying.

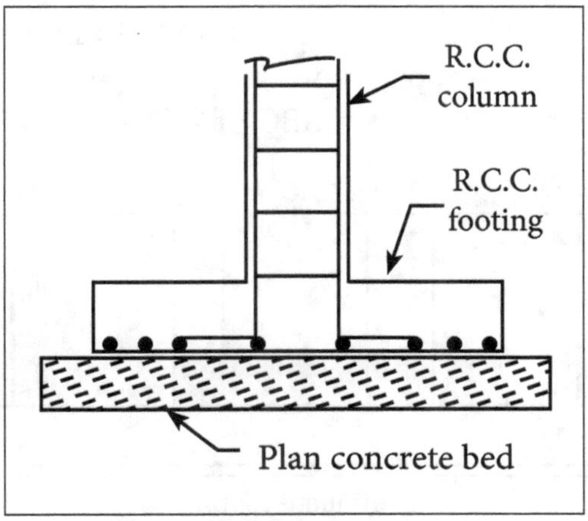

Footing with uniform thickness.

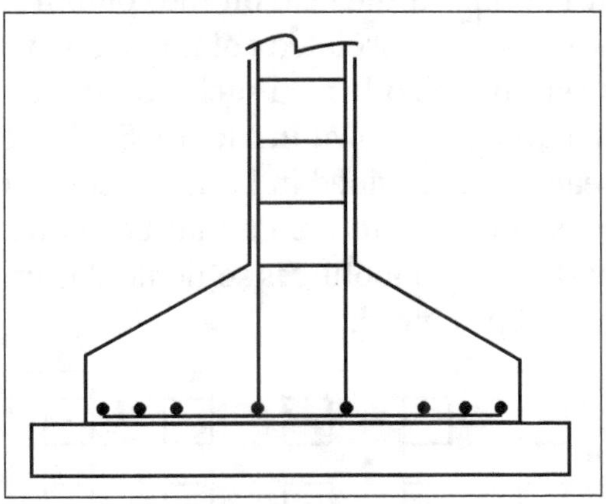

(b) Sloping footing
Isolated R.C.C. Footing.

- Combined Footings: Common footings are provided for two columns. This type of footing is necessary when a column is very close to the boundary of the property and hence there is no scope to project footing much beyond the column face.

Figure illustrates a typical combined footing. The footing is to be designed for transferring loads from both columns safely to the soil. The two columns may or may not be connected by a strap beam.

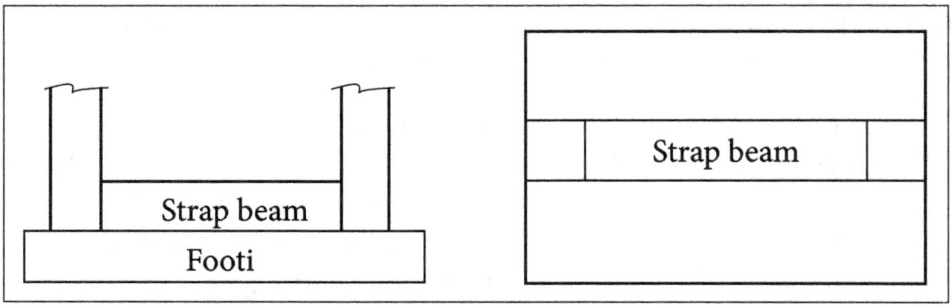

Combined footing [Strap beam may or may not be provided].

- Continuous Footings: If a footing is common to more than two columns in a row it is known as continuous footing. This type of footing is necessary if the columns in a row are closer or if SBC of soil is low. Figure shows this type of footing.

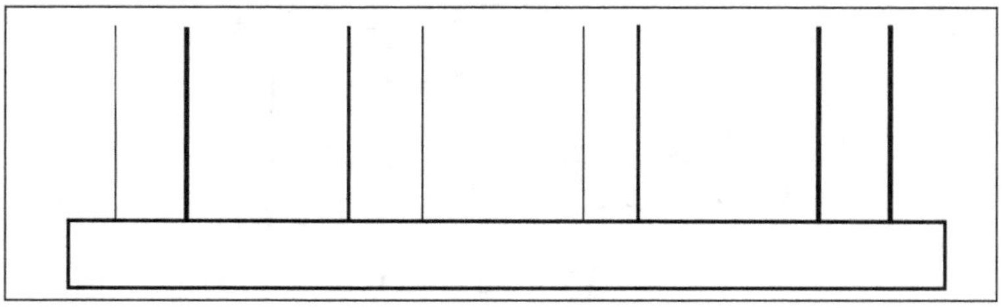

Continuous footing.

- Mat Footing/Raft Footing: If the load on the column is quite high or when the SBC of soil is low the sizes of isolated columns may work out to be to such an extent that they overlap each other. In such situation, a common footing is provided to several columns as shown in Figure. Such footings are known as raft footings. If the beams are provided in both directions over the footing slab for connecting columns the raft foundations may be known as grid foundation also. The added advantage of such footing is settlement is uniform and hence unnecessary stresses are not produced.

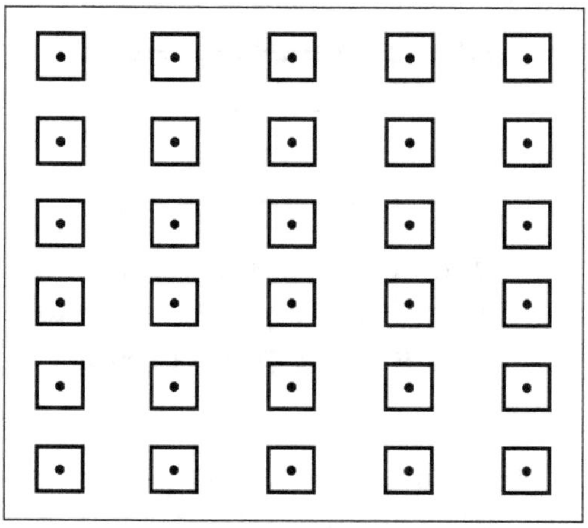

Raft foundation.

- Grillage footing: High rise buildings are built with steel columns encased in concrete. Such columns carry very heavy load and hence they need special foundations to spread the load to a larger area of soil. Grillage foundation is one such special foundation. It consists of one tier or more tiers of I-sections steel beams. Figure illustrates a typical two tier grillage foundation.

Top tier consists of less number but large size steel section while lower tier consists of larger number but smaller size steel sections. Column load is transferred to the top tier through a base plate. The grillage beams are unpainted and are encased in concrete with minimum cover of 100 mm beyond the edges of steel sections.

A minimum clear space of 75 mm must be maintained between the flanges of adjacent grillage beams so that concreting will be made properly. To maintain spacing, pipe separators are used.

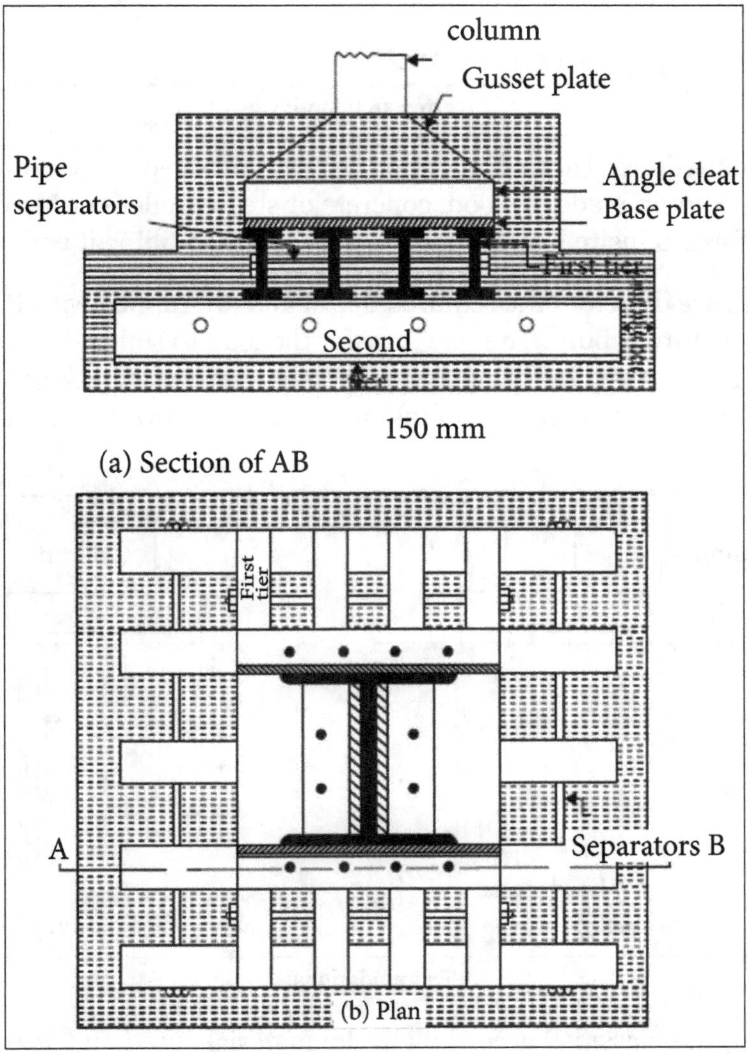

Grillage foundation.

- Arch foundation: Inverted arch foundations are provided in the places where the SBC of the soil is very poor and the load of the structure is through walls. In such cases inverted arches are constructed between the walls. End walls should be sufficiently thick and strong to withstand the outward horizontal thrust due to arch action. The outer walls are provided with buttress walls to strengthen them. Figure shows a typical inverted arch footing.

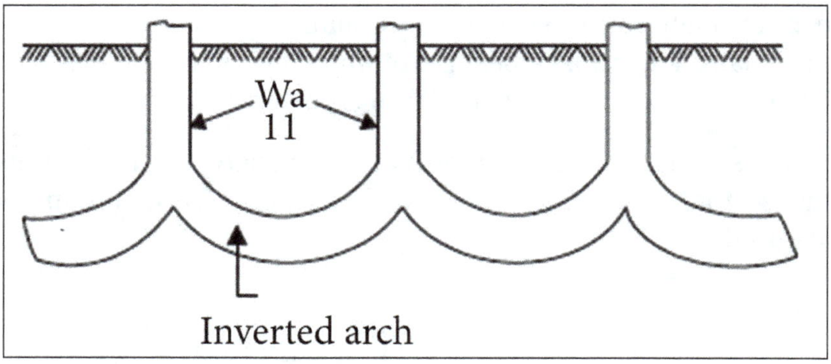

Inverted arch footing.

- Pile foundations: These foundations are called deep foundations. A pile is a slender column made of wood, concrete or steel. A pile is either driven into the soil or formed insitu by excavating a hole and then filling it with concrete.

A group of piles are driven to the required depth and are capped with R.C.C. slab over which super structure is built. The pile transfer the load to soil by friction or by direct bearing in the latter case piles being taken up to hard strata. This kind of foundations is used when top soil is not capable of taking the load of the structure even at 3 - 4 m depth.

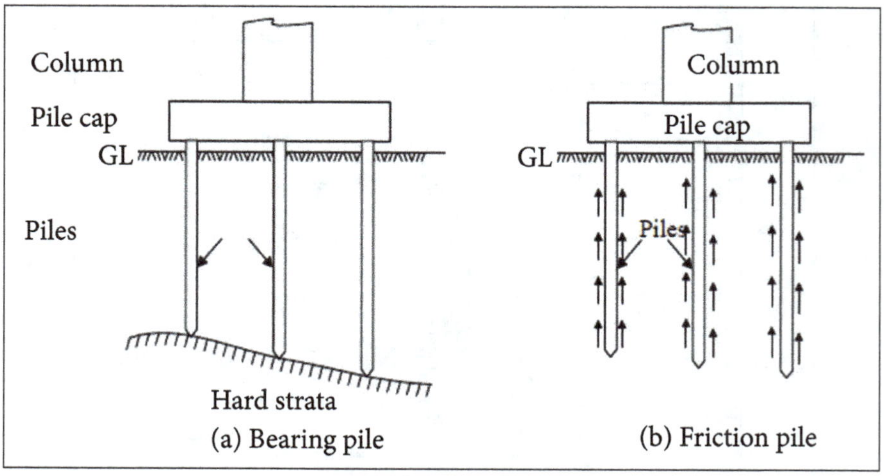

Pile foundations.

Pile foundations are classified according to the materials used and also on the nature of load transfer.

Piles are classified as:

1. Timber Piles

Circular seasoned wood will be used as piles. Their diameter may vary from 200 mm to 400 mm. Similarly square piles of sizes 200 mm to 400 mm are also used. The length of timber pile should not be more than 20 times its lateral dimension.

The bottom of the pile is sharpened and is provided with iron shoe, so that it can be driven in the ground easily by hammering. These piles should be always kept below water table otherwise alternating wet and dry condition cause the decay. These piles are cheap and will be easily driven rapidly. The main disadvantage is their load carrying capacity is low and are likely to be damaged during driving in the soil.

2. Concrete Piles

These piles may be further classified as precast piles and cast in-situ piles. Precast piles are reinforced with steel and are manufactured in factories. The cross-section dimension varies from 200 mm to 500 mm. Square, circular and octagonal sections are commonly used.

The length of piles may be up to 20 m. They are provided with steel shoe at the lowest end. These piles will carry fairly large loads. These piles are highly resistant to biological and chemical actions of the soil. The disadvantage of these piles is they require more time to manufacture and are heavy to handle. Figure (a) and (b) show concrete piles.

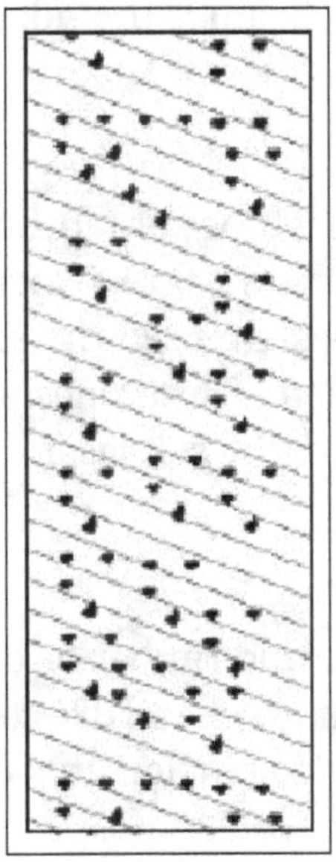

Pressure pile.

Cast insitu concrete piles are formed first by boring the holes in the soil and then concreting them. Concreting is usually made using casing tubes. If the hole is filled with only plain concrete it is pressure pile. The load carrying capacity of the piles are increased by providing enlarged base.

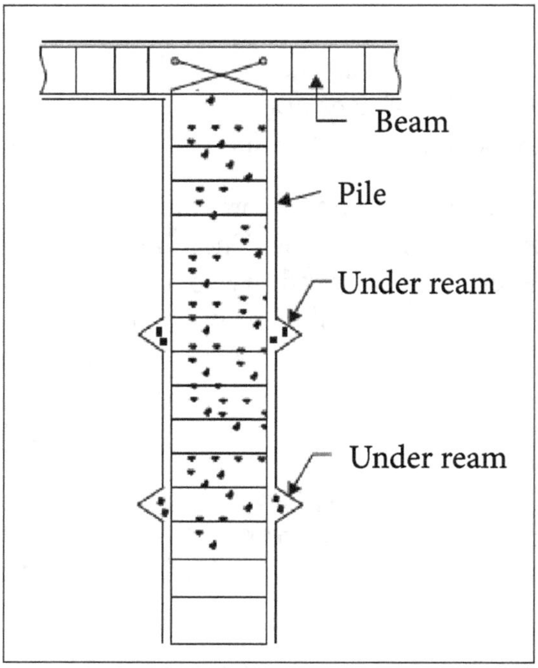

The reinforcement caging may be inserted in the bored holes and to increase load carrying capacity one or two under reams may be formed. After that concreting may be carried out. Such piles are called under reamed piles. These piles are provided at regular interval of 2 to 4 m and capping beam is provided over them.

3. Steel Piles

A steel pile is a rolled steel I sections, tubes or fabricated in the form of box. These piles are mostly used as bearing piles since surface available for friction is less and also the coefficient of friction is less. If tubes are used the soil inside the tube is driven out by compressed air and concrete is filled. These piles are very useful for driving close to existing structures since they disturb the soil least.

4. Composite Piles

Composite piles may be of concrete and timber or of concrete and steel. Wooden piles should not be subjected to alternating wet and dry conditions. Hence they are preferred for the portion below water table. The portion above water table are built with cast insitu concrete piles.

If the required length of steel piles is less than the depth of pile many times upper portions are built with concrete. Thus steel and concrete composite piles are sometimes used.

Classification of Piles According to Load Transfer

According to the load transfer to the soil piles may be classified as:

- Bearing piles rest on hard strata and transfer the load by bearing. Such piles are preferred. These piles are used if the hard strata are available at reasonable depth.
- Friction piles transfer the load to the soil by the friction between soil and the pile. Such piles are used if hard strata are not available to a considerable depth. The friction developed is to be properly assessed before deciding the length of the pile. The surface of such piles is made rough to increase the skin friction so that required length of pile is reduced.

1.5 Brick Masonry Walls

Brick masonry is built with bricks bonded together with mortar. For temporary sheds mud mortar may be used but for all permanent buildings lime or cement mortars are used.

The various kinds of bonds generally used in brick masonry are:

- Stretcher Bond: A stretcher is the longer face of the brick as seen in the elevation. In the brick of size 190 mm × 90 mm × 90 mm, 190 mm × 90 mm face is the stretcher. In stretcher bond masonry all the bricks are arranged in stretcher courses as shown in figure. But care should be taken to break vertical joints. This kind of construction is useful for the construction half brick thick partition wall.

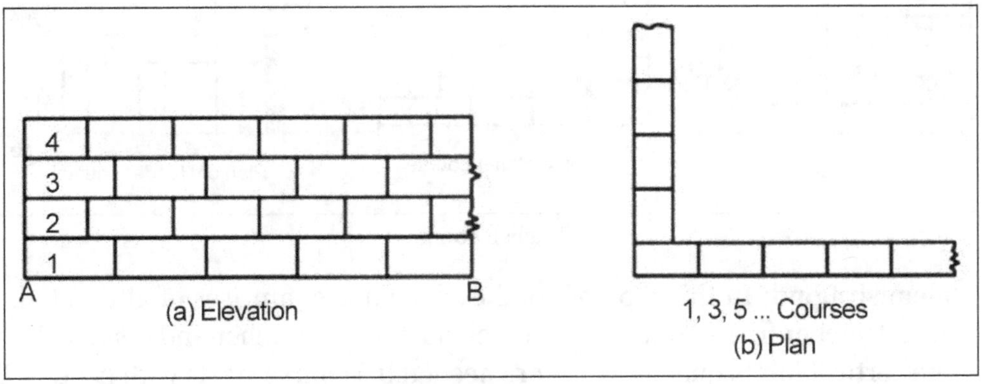

Stretcher bond.

- Header Bond: A header is the shortest face of the brick as seen in the elevation. In a standard brick it is 90 mm × 90 mm face. In header bond brick masonry all the bricks are arranged in the header courses as shown in figure. This kind of bond is useful for the construction of one brick thick wall.

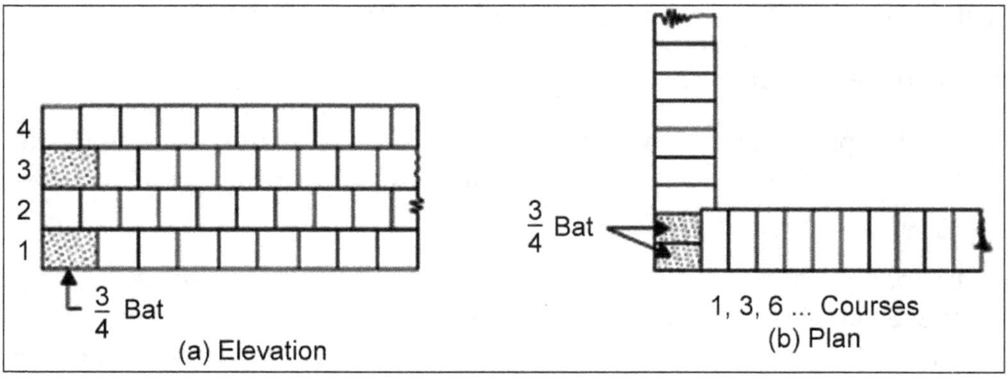

Header bond.

- English Bond: In this alternate courses consist of headers and stretchers. This is considered to be the strongest bond. Hence it is normally used bond for the walls of all thicknesses. To break continuity of vertical joints a brick is cut lengthwise into two halves and used in the beginning and end of a wall after first header. This is known as queen closer. Figure shows typical one brick and one and half brick thick wall with English bond.

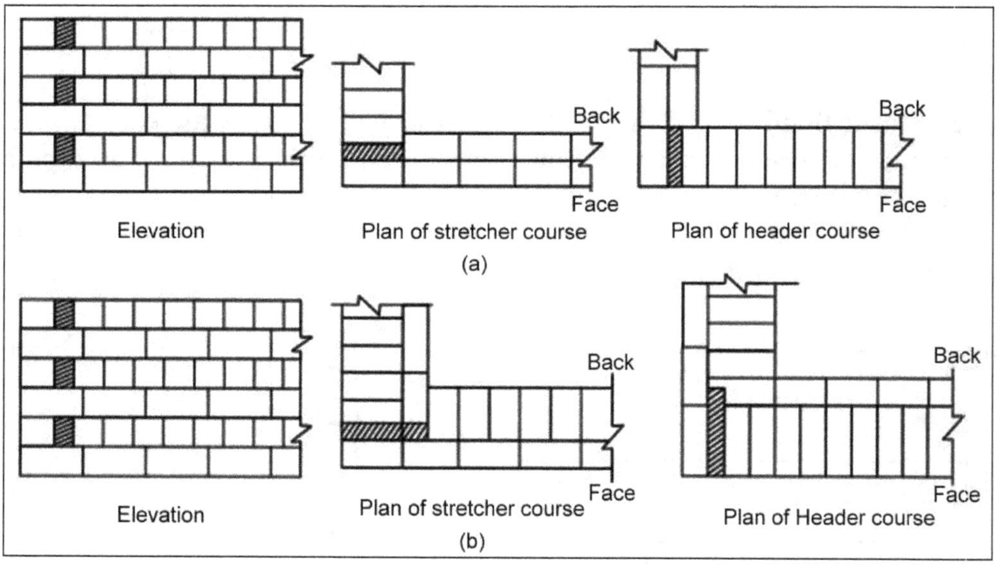

English bond.

- Flemish Bond: In this type of bond each course comprises of alternate header and stretcher figure. Alternate courses start with stretcher and header. To break the vertical joints queen closers are needed if a course starts with header. Every header is centrally supported on the stretcher below it. Flemish bonds are further classified as:
  - Double Flemish Bond.
  - Single Flemish Bond.

In case of double flemish bond both faces of the wall have flemish look i.e. Each course consist of alternate header and stretcher whereas single flemish bond outer faces of walls have flemish look whereas inner faces have look of English bond. Construction of flemish bond needs greater skill. It provides more pleasing appearance. But it is not as strong as English bond. If only pointing is to be used for finished wall flemish bond may be used to get good aesthetic view. If plastering is going to be used it is better to use English bond.

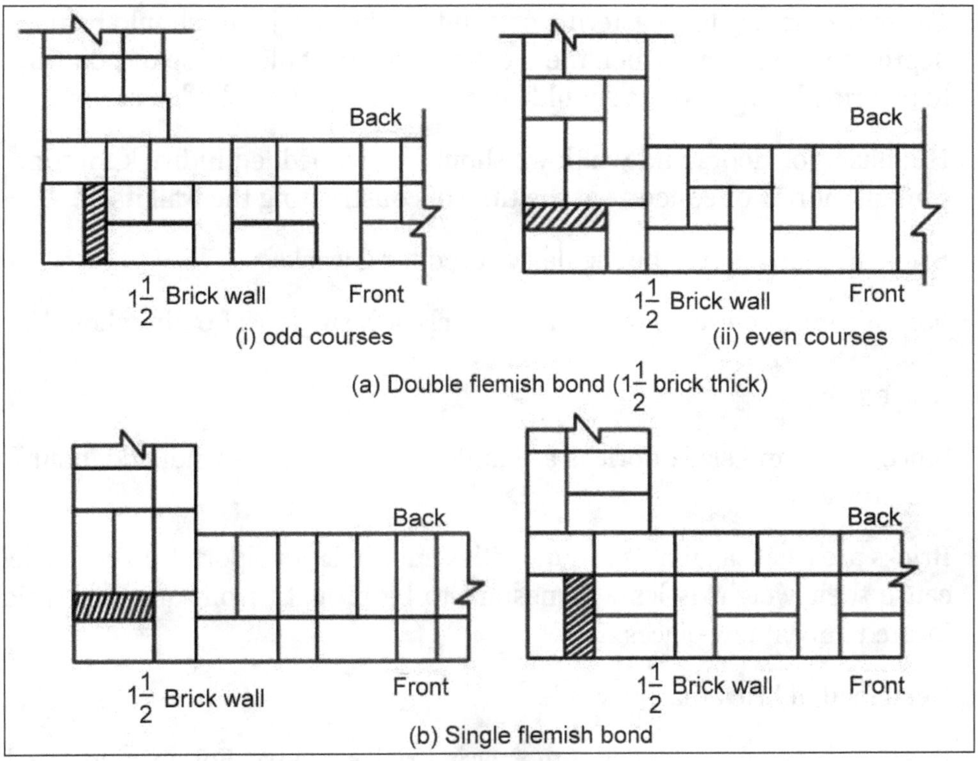

Flemish bond.

The following points should be observed in the construction of brick masonry:

- Use bricks of good quality with uniform colour, well burnt with exact shape and size.
- Before using the bricks in masonry they should be soaked in water for 2 hours so that bricks do not absorb water from the mortar.
- Bricks should be laid with the frog pointing upward.
- Construction of brick wall should start from the end or corner.
- Brick courses should be perfectly horizontal.
- Verticality of the wall should be ensured by frequently checking with plumb-bob.

- Mortar used should be as per specification.
- Whenever work is stopped brick masonry should be left with toothed end.
- Use of brick bats should be avoided.
- Walls should be raised uniformly. In case difference between adjoining walls can be more than 1 m. In a day no wall should be raised by more than 1.5 m.
- To get proper key for plastering or pointing the face joints should be raised to a depth of 12 to 20 mm when the mortar is green. If plastering or pointing is not to be provided face joints should be stuck flush and finished neatly.
- Holdfasts for doors and windows should be embedded in brick masonry with cement mortar or concrete at the time of constructing the wall itself.
- Brick masonry should be regularly cured for 2 weeks.
- For carrying out brick work at higher levels only single scaffolding should be used.

Advantages

- Since shape and size of bricks are uniform it do not need skilled labour for the construction.
- Bricks are easily available around cities and their transportation cost is less because their weight is less. Stones are to be brought from quarries which are located only at few places.
- Dead load of brick masonry is less.
- It is possible to use all types of mortar in brick masonry. For unimportant buildings even mud mortar can be used.
- Bricks are light in weight and hence handling them is easy.
- Thinner walls can be constructed with bricks but it is not so with stones.
- Brick masonry has better fire and weather resistance compared to stone masonry.
- It is easy to form openings for doors and windows.
- In brick masonry mortar joints are thin and hence construction cost is reduced considerably.

Disadvantages

- Strength of brick masonry is less than that of stone masonry.

- Brick masonry needs plastering and plastered surface needs colour washing whereas stone masonry don't need them and hence maintenance cost is more in brick masonry.
- Durability of brick masonry is less.
- Brick masonry absorbs water and there are possibility of dampness. There is no such problem in stone masonry.
- Stone masonry gives massive appearance and hence monumental buildings are built in stone masonry.
- More architectural effects can be given in stone masonry compared to that in brick masonry.

1.6 Pointing and Plastering

The exposed surface of walls are to be provided with pointing or plastering. The method of pointing is adopted for stone masonry or brick masonry while plastering is carried out for exposed surfaces of sailings, walls, columns, etc.

Objects of Pointing and Plastering

- To improve the appearance of the structure as a whole and to give smooth surface.
- To protect the exposed surfaces from the effects of atmospheric action.
- To rectify the defective workmanship or to conceal inferior materials.

1.6.1 Pointing

The term pointing used to denote finishing of mortar joints of either stone masonry or brick masonry. The joints are raked out to a depth of about 20 mm and then, these spaces are filled up by suitable mortar in the desired shape.

It is desirable to avoid pointing as far as possible. This is due to the fact that pointing involves raking out of joints which are constructed with good mortar and filling the joints with mortar which, in several cases, is not sufficiently watered. For this reason, the pointing work of new structure should follow the masonry work in progress. The joints are thus raked out when the mortar has not set.

Mortar for Pointing

The pointing is generally adopted for the finishing of exposed external walls of a structure. It is cheap in the first coat, but it requires frequent replacement. The pointing is carried out either in lime or in cement mortar. The lime mortar consist of equal volumes of lime and sand. These two materials are carefully mound in a mortar mill. The sand to be used for preparing lime mortar should be clean, fine and free from any organic impurities.

The cement mortar consists of equal volume of cement and sand. The cement should comply with standard requirements and sand should be clean, fine and free from organic impurities. The materials are thoroughly mixed in dry condition before water is added to them. The mixing of materials is done on watertight platform and mortar of one cement bag only is prepared at a time and this quantity of mortar is consumed within 30 minutes after adding water.

Method of Pointing

The pointing is carried out as follows:

- The mortar of masonry joints to be covered by pointing is raker out at least to a depth of 20 mm.
- The dust from the masonry joints is removed by the brushes.
- The surface is then washed within clean water and it is kept wet for a few hours.
- The mortar is then carefully placed in desired shapes in these prep are joints.
- The mortar is placed by a small towel and it is slightly pressed to bring into close contact with the old interior mortar of the joint.
- The finished surface is well-watered for a period of at least 3 days, It lime mortar is used and 10 days, if cement mortar is used.

Types of Pointing

- Beaded pointing.
- Flush pointing.
- Recessed pointing.
- Rubbed or Keyed or grooved pointing.
- Struck pointing.
- Tuck pointing.
- Vee-pointing.

1.6.2 Plastering

The term plastering is used to describe the thin plastic covering that is applied on the surfaces of walls and ceilings. The plastering removes the unevenness of the surfaces and sometimes the plastering is used to develop decorative effects.

Objective of Plastering

- To conceal defective workmanship.
- To give smooth surface to avoid catching of dust.
- To give good look.
- To protect the wall from rain water and other atmospheric agencies.
- To protect surfaces against vermin.

Requirements of Good Plasters

- It should adhere to the background and should remain adhered during all variations of the climatic changes.
- It should be cheap and economical.
- It should be hard and durable.
- It should be possible to apply it during all weather conditions.
- It should effectively check the entry or presentation of moisture from the surface.
- It should possess good workability.

Mortar for Plastering

The selection of types of mortar for plaster depends on various factors like availability of building materials, atmospheric conditions, durability required, finishing desired, locations of the surface, etc. There are mainly three kinds of mortar which can be used for the process of plastering:

- Lime mortar: The lime mortar consists of equal volume of lime and sand, and these two materials are carefully ground in mortar mill. The fat lime is recommended for plaster work because the fat lime contains 75% of CaO and it combines with CO_2 of atmosphere and gives $CaCo_3$ quickly. Thus, the limes set quickly, however it imparts low strength and hence, it can be used only for plaster work. The hydro lick lime contains practical which slakes slowly as they come in contact with atmosphere. This result into blisters on the surface known as the blowing. The sand to be used for preparing lime mortar for plastering work should be clean, coarse and free from any organic impurities.

In order to improve the binding properties of mortar, a type of fragrant gum-gugal is added to the lime mortar. The quantity of gugal is about 16N per m^3 of mortar. The gugal with enough quantity of fresh water before it is with the lime mortar. The quantity of hump is about 10 N per m^3 of mortar. The mortar thus prepared is kept for two days before use and it is turned ones in a day to make the mortar of uniform nature.

- Cement mortar: The cement mortar consist of one part of cement to four parts of clean, coarse and angular river sand by volume. The materials are thoroughly mixed in dry condition before water is added to them. The mixing of materials is done on a watertight platform and mortar of one cement bag only its prepared at a time and this quantity of mortar is consumed within 30 minutes after adding water.
- Water-proof mortar: This mortar is water proof and is prepared by mixing one part of cement, two parts of sand and pulverized alum at at the rate of 120 N per m^3 of sand. In the water to be used, 0.75 N of soft soap is dissolved per one liter of water and this soap water is the added to the dry mix.

1.7 Flooring

Purpose of flooring is to get a good hard level and beautiful surface for living. The floors directly resting on the ground are known as ground floors while the floors of each storey are called upper floors.

Ground Floor: Apart from giving good finished surface these floors should have good damp resistance. The ground surface is rammed well and a layer of red earth or sand is placed which is compacted. A layer of broken bricks, stones etc. is provided up to 150 mm below floor finish level and rammed. While ramming the surface is kept moist to get good compaction. Then 1 : 4 : 8 concrete of 100 to 150 mm thickness is provided as base course. Over this bed floor finish is laid. The kind of flooring used are:

- Mud and Moorum Flooring: These floorings are used in low cost housing specially in villages. Over the hard layer of earth filling mud or moorum layer is provided. The floor requires a thin wash of cow dung at least once a weak.
- Brick Flooring: This is also a cheap floor construction. It is usually used in go down sand factories. Bricks are laid flat or on edges. Bricks of good quality must be used for the construction. Brick layer is provided on sand bed or on lean

concrete (1: 8 : 16) bed. In both cases joints are rendered flush and finished with cement mortar.

- Flag Stone Flooring: Laminated sand stones or slates of 20 mm to 40 mm thick in the form of slabs of 300 mm × 300 mm or 450 mm × 450 mm or in the form of rectangles of size 450 mm × 600 mm are used as floor finishes. The stone slabs are laid on 20 to 25 mm thick mortar spread over concrete bed. The joints are to be finished with rich mortar.

- Cement Concrete Floors: It is modestly cheap and durable floor and hence normally used in residential, commercial and industrial buildings. It consists of two courses-base course and wearing coat. Base course is laid over well compacted soil. Its thickness is typically 75 mm to 100 mm. It consists of lean cement concrete mix (1 : 4 : 8) or lime concrete containing 40% of 1 : 2 lime mortar and 60% of coarse aggregate of 40 mm size.

  After base coarse is hardened wearing coat of 40 mm is laid. It consists of panels of 1 m × 1m, 2 m × 2 m or 1 m× 2 m. Alternate panels are laid with 1: 2: 4 concrete using wooden, glass or asbestos strip separators of 1.5 mm to 2.0 mm thickness.

  To get good bond between base coarse and wearing coat cement slurry wash is given before laying wearing coat panels. After 3 to 4 days of laying of one set of panel another alternate panels are laid. Top of these panels are finished by tamping the surface with wooden floats and tapping with trowels till cement slurry appears on top. It requires curing for 7 to 14days. To get good appearance many times red-oxide finishing coat is provided.

- Terrazzo Flooring: Terrazzo finishing coat is applied over concrete flooring to get pleasing appearance. Terrazo finish consists of 75 to 80% of surface marble chips embedded in cement mortar.

  Marble chips are mixed in cement in the proportion 1: 1.25 to 1 : 2 and about 6 mm terrazo topping is laid. The top is tamped and rolled. Also marble chips are spread during tamping to get proper distribution of marble chips on the surface. After drying it for 12 to 20 hours, it is cured for 2 to 3 days.

- Mosaic Flooring: This consists of a finishing coat of small pieces of broken tiles of China glazed or of marble arranged in different patterns set in lime-surkhi or cement mortar. The base coarse is concrete flooring and on it 30 to 40 mm mortar layer is provided. On this mortar layer broken pieces of China glazed or marble are set to get various attractive patterns. After 20 to 24 hours of drying the top is rubbed with carborundum stone to get smooth and polished surface.

- Marble Flooring: Marble slabs are cut to get marble tiles of 20 to 25 mm thickness. They are laid on floors similar to other tiles. With power driven machine surface is polished to get even and shining surface. This kind of flooring is widely used in hospitals and temples.

- Tiled Flooring: This is an alternative to terrazo flooring used usually used in residential office and commercial buildings. Tiles of clay, cement or terrazo of standard sizes are manufactured in factories under controlled conditions. On the concrete base, 25 mm to 30 mm thick mortar is laid and these tiles are placed and pressed with trowel or wooden mallet. Before placing tiles care is taken to see that neat cement slurry is applied to bottom side and sides of tiles to get good bond. Next day joints are cleaned of loose mortar and raked up to 5 mm depth. Then which is filled with coloured cement slurry to get uniform colour on the top surface. After curing for 7 days grinding and polishing is made as in the case of terrazo flooring.
- Timber Flooring: Timber flooring are used in dancing halls and in auditoriums. Timber plates are directly placed on concrete bed or may be provided over timber frame work. In latter case it is necessary to give proper ventilation below the floor. This flooring is costly.
- Rubber Flooring: Tiles or sheets of rubber with fillers such as cotton fibres, asbestos fibre or granulated cork are manufactured in variety of patterns and colours. These sheets or tiles are fixed to concrete or timber floors. These floors are attractive and noise proof. However they are costly.
- P.V.C. Flooring: Poly-Vinyl-Chloride (PVC) is a plastic which is available in different colour and shade. Nowadays tiles of this material are used widely. Adhesives are applied on concrete base as well as on bottom of PVC tiles. Then the tile is pressed gently with 5 kg wooden roller till the oozing of adhesive is seen. The oozed out adhesive is wiped and the floor is washed with warm soap water. The floor finish is smooth, attractive and will be easily cleaned. However it is slippery and costly.

1.7.1 Roofing

Roof is the upper most portion of the building which protects the building from rain, wind and sun. Various types of roofs are classified into three types:

1. Flat Roofs

Flat roofs are used in plains where rainfall is less and climate is moderate. Pitched roofs are preferred wherever rainfall is more. Shells and folded plate roofs are used to cover large column, free areas required for auditoriums, factories etc. Brief description of these roofs is presented below:

These roofs are nearly flat. But slight slope (not more than 10°) is given to drain out the rain water. All types of upper storey floors will serve as flat roofs.

Several times top of these roofs are treated with water proofing materials like mixing water proofing chemicals in concrete providing ciba concrete. With advent of reliable water proofing techniques such roofs are constructed even in areas with heavy rain fall.

The advantages of flat roofs are:

- The roof will be used as a terrace for playing and celebrating functions.
- At any latter stage the roof will be converted as a floor by adding another storey.
- They will suit to any shape of the building.
- Over-head water tanks and other services will be located easily.
- They will be made fire proof easily compared to pitched roof.

The disadvantages of flat roofs are:

- They cannot cover large column free areas.
- The dead weight of flat roofs is more.
- Leakage problem may occur at latter date also due to development of cracks. Once leakage problem starts it needs costly treatments.
- The initial cost of construction is more.
- In places of snow fall flat roofs are to be avoided to reduce snow load.
- Speed of construction of flat roofs is less.

2. Pitched Roofs

In the areas of heavy rain falls and snow fall sloping roof are used. The slope of roof shall be more than 10°. They may have slopes as much as 45° to 60° also. The sloped roofs are called pitched roofs.

The sloping roofs are preferred in large spanned structures like workshops, factory buildings and ware houses. In all these roofs covering sheets such as A.C. sheet, G.I. sheets, tiles, slates etc. are supported on suitable structures.

The pitched roofs are classified into:

- Single Roof: If the span of roof is less than 5 m the following kinds of single roofs are used:
    - Lean to roofs.
    - Coupled roofs.
    - Coupled-close roof.

- Collar beam roof.

In all these roofs rafters placed at 600 mm to 800 mm spacing are main members taking load of the roof. Battens run over the rafters to support tiles. Figure shows various kinds of single roofs:

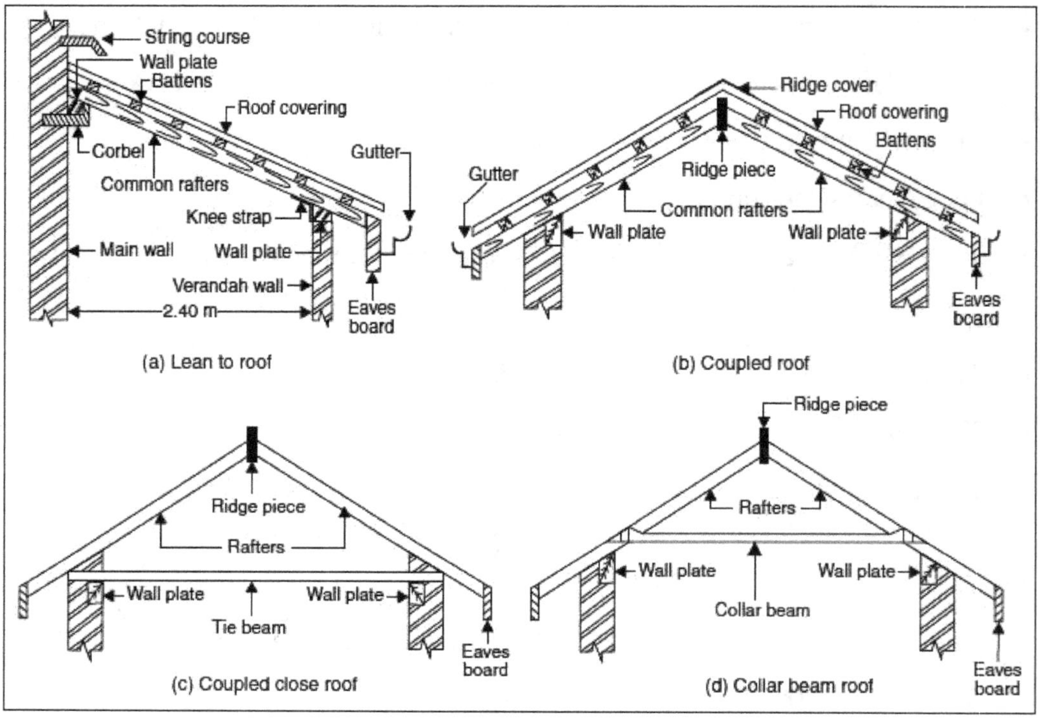

Single roofs.

- Double or Purlin Roofs: If span exceeds the cost of rafters increase and single roof becomes uneconomical. For spans more than 5 m double purlin roofs are preferred. The intermediate support is given to rafters by purlins supported over collar beams. Figure illustrated a typical double or purlin roof:

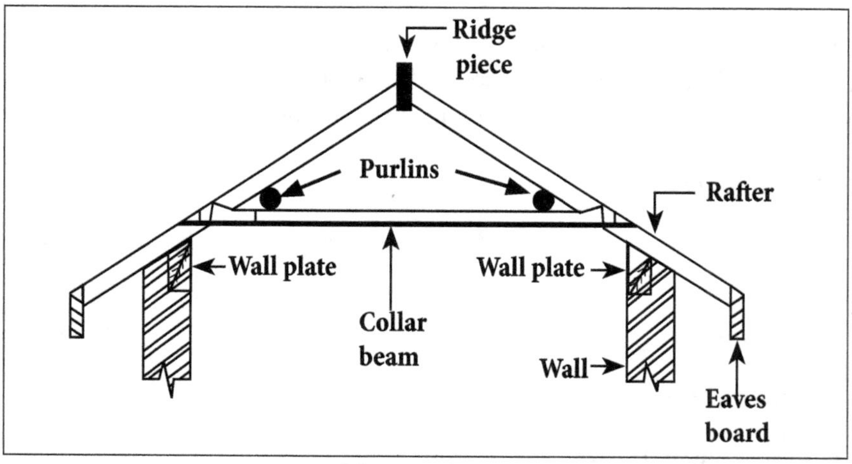

Double or purlins roofs.

- Trussed Roof: If span is more a frame work of slender members are used to support sloping roofs. These frames are known as trusses. A number of trusses are placed lengthwise to get wall free longer halls. Purlins are provided over the trusses which in turn support roof sheets. For spans up to 9 m wooden trusses are used but for larger spans steel trusses are a must.

In case of wooden trusses suitable carpentry joints are made to connect various members at a joint. Bolts and straps are additionally used. In case of steel trusses joints are made using gusset plates and by providing bolts or rivets or welding.

Depending upon the span, trusses of various shapes are used. End of trusses are supported on walls or on column. Figure illustrated different shapes of trusses used:

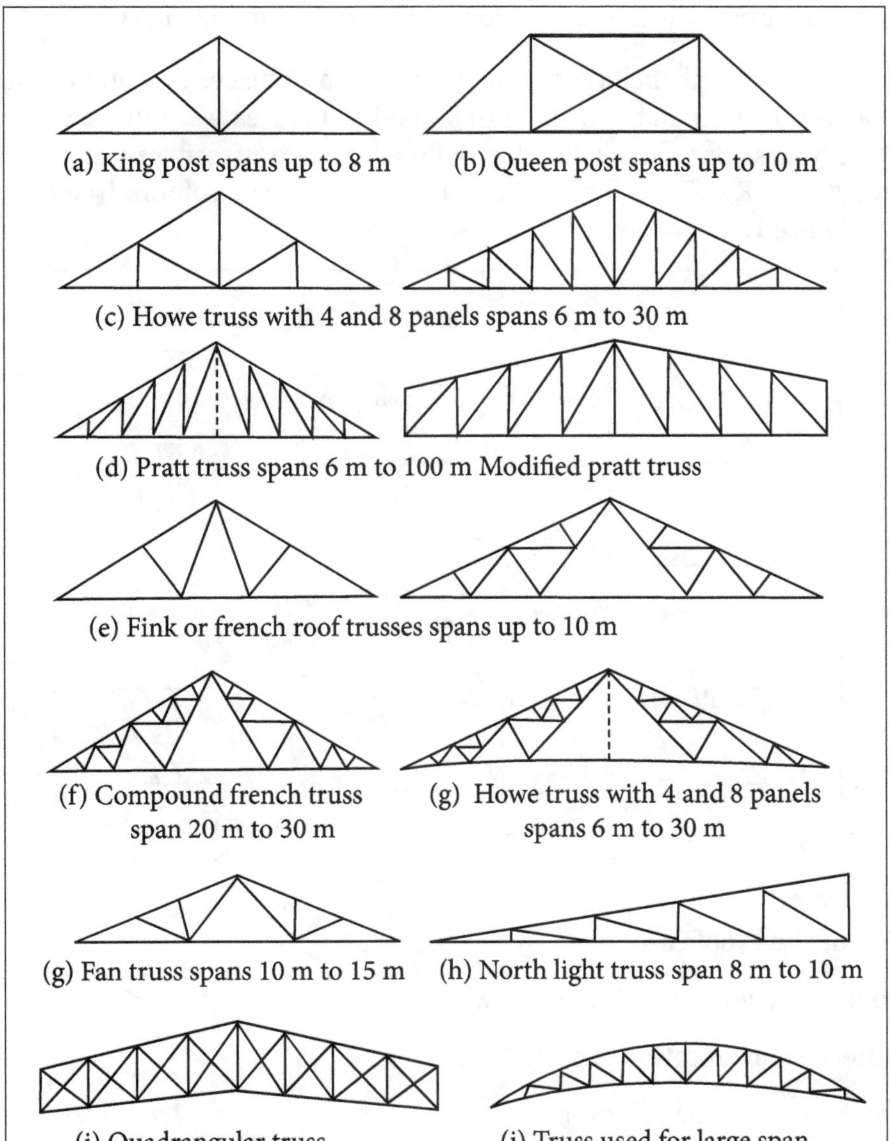

(a) King post spans up to 8 m
(b) Queen post spans up to 10 m
(c) Howe truss with 4 and 8 panels spans 6 m to 30 m
(d) Pratt truss spans 6 m to 100 m Modified pratt truss
(e) Fink or french roof trusses spans up to 10 m
(f) Compound french truss span 20 m to 30 m
(g) Howe truss with 4 and 8 panels spans 6 m to 30 m
(g) Fan truss spans 10 m to 15 m
(h) North light truss span 8 m to 10 m
(i) Quadrangular truss
(j) Truss used for large span

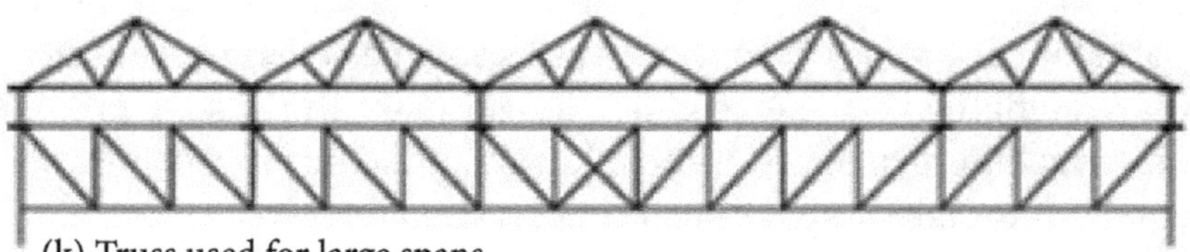

(k) Truss used for large spans

Types of trusses.

3. Shells and Folded Plate Roofs

Shell roof is defined as a curved surface the thickness of which is small compared to the other dimensions. In these roofs lot of load is transferred by membrane compression instead of by bending as in the case of conventional slab and beam constructions.

Caves are having natural shell roofs. An examination of places of worships built in India, Europe and Islamic nations show that shell structures were in usage for the last 800 to 1000 years. But the shells of middle ages were massive masonry structures but nowadays thin R.C.C. shell roofs are built to cover large column free areas. Figure shows usually used shell roofs:

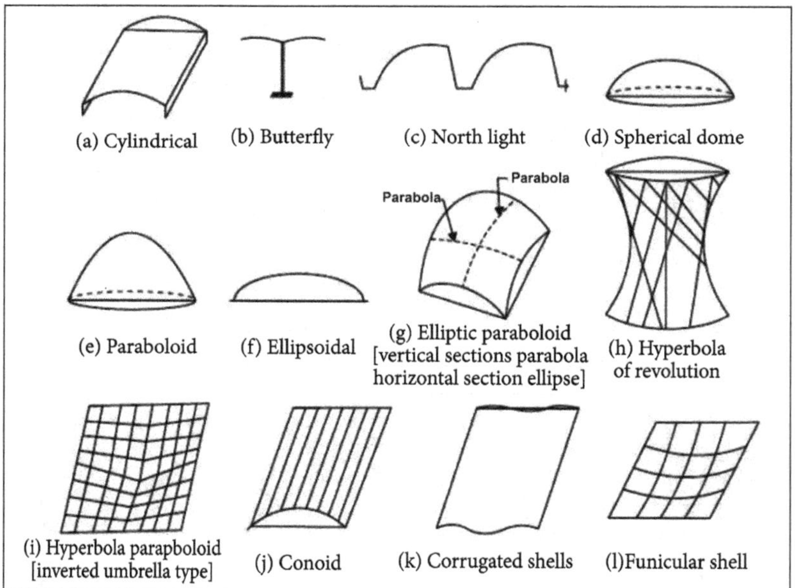

Types of shell roof.

Advantages of shell roofs are:

- Good from aesthetic point of view.
- Material consumption is quite less.
- Form work will be removed early.
- Large column free areas can be covered.

Disadvantages:

- Top surface is curved and hence advantage of terrace is lost.
- Form work is costly.

Folded plate roofs are looked as slab with a number of folds. These roofs are also known as hipped plates, prismatic shells and flatwork. In these structures bending is reduced and lots of load gets transferred as membrane compression. But folded plates are not so efficient as shells. Figure shows typical folded plate roofs:

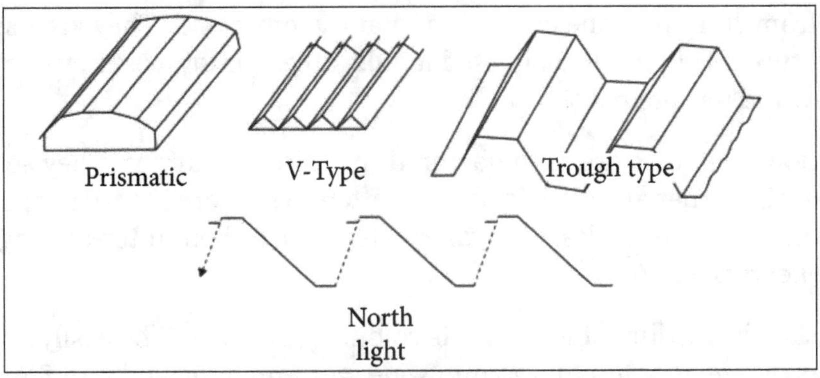

Types of folded plate roofs.

Advantages:

- Form work required is relatively simpler.
- Movable form work will be employed.
- Design involves simpler calculations.

Disadvantages:

- Folded plate consumes more material than shells.
- Form work will be removed after 7 days while in case of shells it can be little earlier.

Roof Coverings for Pitched Roofs

Different types of covering materials are available for pitched roofs and their selection depends upon the climatic conditions, fabrication facility and availability of materials and affordability of the owner. Usually used pitched roof covering materials are:

1. Thatch Covering: These coverings are provided for small spans mainly for residential buildings in villages. Thatch is a roof covering of straw, reeds or similar materials. The thatch is well-soaked in water or fire resisting solution and packed bundles are laid with their butt ends pointing towards eves. Thickness differ from 150 mm to 300 mm. They are tied with ropes or twines to supporting structures.

The supporting structure consists of round bamboo rafters spaced at 200 mm to 300 mm over which split bamboos laid at right angles at close spacing. It is claimed that reed thatch will last 50 to 60 years while straw thatch may last for 20 to 25 years.

The advantage of thatch roof is they are cheap and do not required skilled workers to build them. The disadvantages are they are very poor fire resistant and harbour rats and other insects.

2. Shingles: Wood shingles are nothing but the split or sawn thin pieces of wood. Their size varies from 300 mm to 400 mm and length from 60 mm to 250 mm. Their thickness varies from 10 mm at one end to 3 mm at the other end. They are nailed to supporting structures. They are usually used in hilly areas for low cost housing. They have very poor fire and termite resistance.

3. Tiles: Various clay tiles are manufactured in various localities. They serve as good covering materials. Tiles are supported over battens which are in turn supported by rafters/trusses etc. Allahabad tiles, Mangalore tiles are excellent inter-locking tiles. They give good appearance also.

4. Slates: A slate is a sedimentary rock. Its colour is gray. It will be easily split into thin sheets. Slates of size 450 mm to 600 mm wide, 300 mm long and 4 to 8 mm thick are used as covering materials of pitched roofs in the areas where slate quarries are nearby. A good slate is hard, tough, durable. They are having rough texture and they give ringing bell like sound when struck. They do not absorb water.

5. A.C. Sheets: Asbestos cement is a material which consists of 15 percent of asbestos fibres evenly distributed and pressed with cement. They are manufactured in sufficiently large size. The width of a A.C. sheet differ from 1.0 to 1.2 m and length from 1.75 to 3.0 m. To get sufficient strength with thin sections they are manufactured with corrugation or with affords. They are fixed to the steel purlins using J-bolts.

The roofing is quite economical, waterproof. However not very good thermal resistant. They are normally used as covering materials in ware houses, godowns or for larger halls. In auditorium etc. if these sheets are used false ceilings are provided to get good thermal resistance.

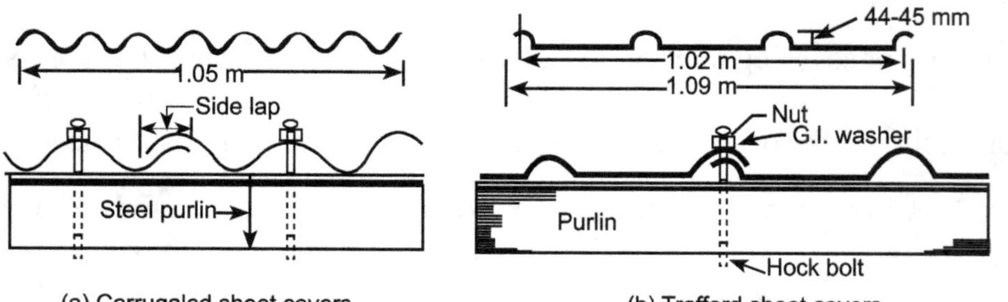

(a) Corrugaled sheet covers

(b) Trafford sheet covers

AC sheet roofing.

6. G.I. Sheets: Galvanized iron corrugated sheets are manufactured in the sizes 1.0 to 1.2 m wide and 1.65 m length. Galvanization of iron makes them rust proof. They are fixed to steel purlins using J-bolts and washers. They are durable, fire proof, light in weight and need no maintenance. They are normally used as covering materials for ware houses, godown, sheds etc.

1.8 Doors, Windows, Lintels and Staircases: Types and their Suitability

1.8.1 Doors

Panelled door: A door created with panels, sometimes formed to pattern, put in between the stiles and rails that form the outside frame of the door.

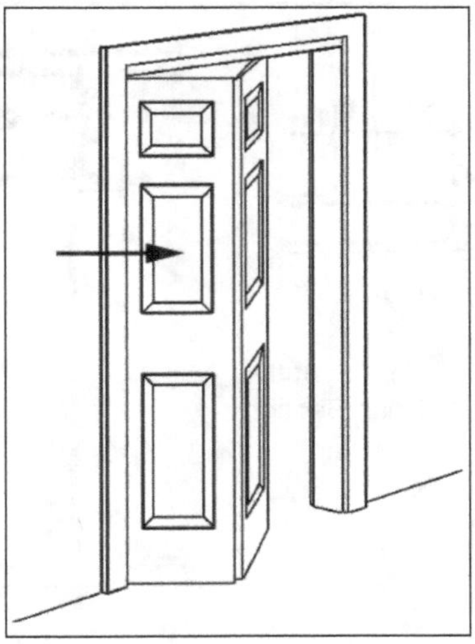

Panel door.

Glassed door: A door unreal while not stiles or rails and consisting entirely of thick, heat-strengthened, or tempered glass.

- This kind is employed in residential and public buildings.
- They will be created totally glazed or part glazed.
- They supplement the natural lighting provided by windows or to create the inside of the space visible from abutting rooms.
- Encase of part glazed, the lowest 1/3rd half is typically decorated and higher 2/3 half is glazed.
- Totally glazed doors area unit suggested wherever enough lightweight is needed through the door openings like in looking malls, entrance halls etc.

Fully Glazed Door.

Partly Glazed Door.

Plan of Door

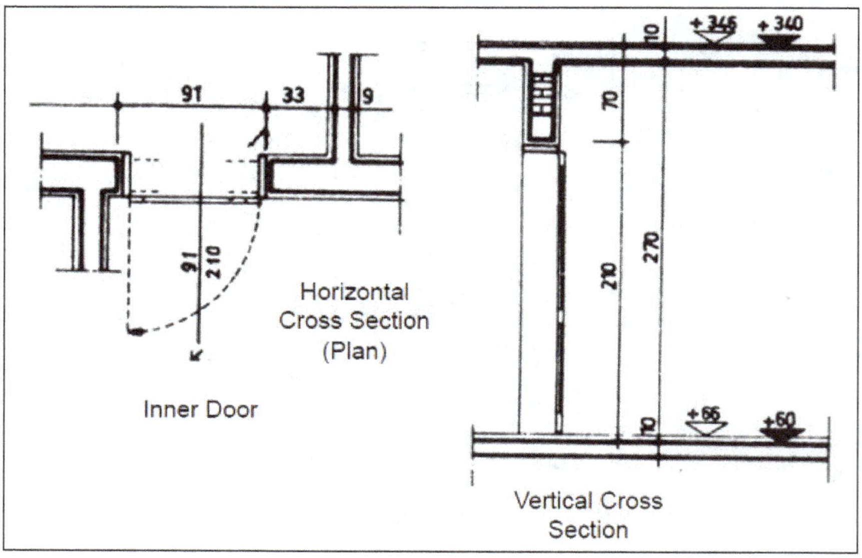

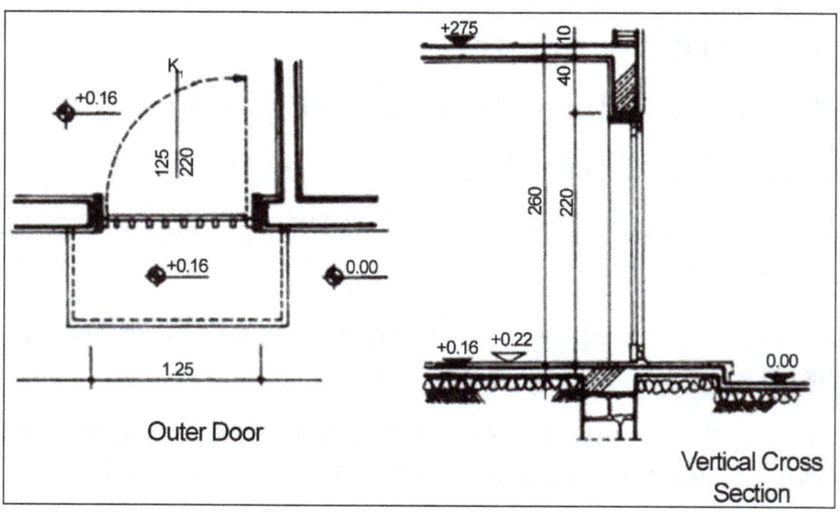

1.8.2 Windows

Glazed window: This is a type of window wherever panels area unit totally glazed. The frame has designs, high rail and a bottom rail. The area between high and bottom rail is split into variety of panels with tiny timber members referred to as sash bars or glazing bars. The glass panel's area unit cut one.5-3.0 millimeter smaller in size than the panel size to allow movement of sash bars. Glass panes area unit mounted to sash bars by putty or by timber beads.

Casement Windows: It is a normal window that is often employed in residential similarly as publically buildings. The shutters of the window open like shutters of the doors. The window includes a frame and shutters consist of designs, top rail, bottom rail, intermediate rails, vertical and horizontal sash bars with glass panels. The panels are also glazed or part glazed.

Fixed windows: These consist of a frame and stationary sash. When employed in conjunction with operable window units. The thickness of the mounted sash ought to approximate the cross sectional dimension of the operative sashes.

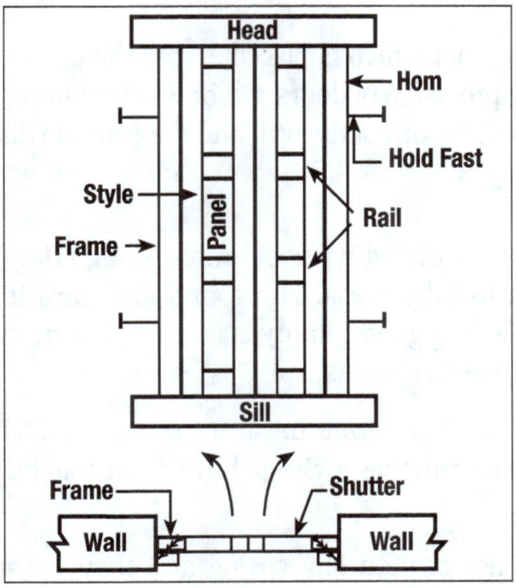

Side hung: Side decorated or window sash windows have operative sashes that area unit facet hinged and typically swing outward. When open the sash is ready to direct ventilation. The inner finish of the sash might slide on the track on the sill or vertical because the sash swings outward. A pair of sashes may close on a mullion or have a floating astragal to close on each other.

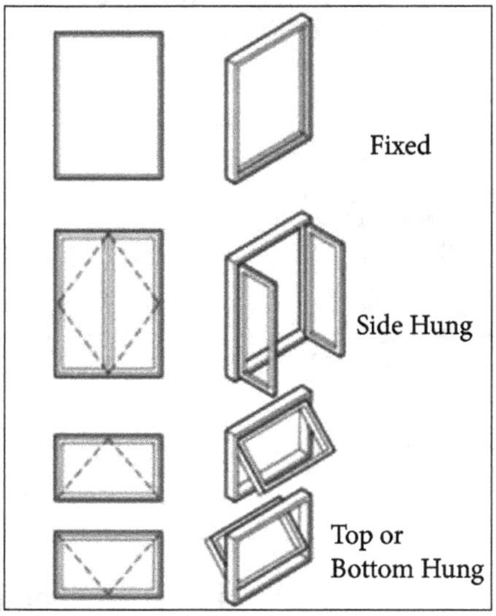

Top or bottom hung: Top decorated windows have operative sashes that swing outward on hinges hooked up to high of their frames. Bottom decorated windows have operative sashes that swing inward on hinges hooked up to the lowest of their frames.

When open the sash is ready to direct ventilation. The sashes is additionally stacked vertically with sashes closing on one another or on meeting stiles.

1.8.3 Lintels

A lintel is a horizontal member which is placed across the openings. Openings are invariably left in the wall for the provision of doors, windows etc. A lintel is thus a sort of beam the width of which is equal to the width of the wall and the ends of which are built into the wall.

Classification:

- Timber lintels are the oldest types of lintels though they are not commonly used nowadays except in hilly areas. The sound and hard it. As the timber is easily liable to catch fire only good quality is constructed over coat of suitable preservative should be used as lintels.
- Stone lintels are used in stone masonry structures. This consists of a simple stone slab of greater thickness. Stone lintels can also be provided over openings in brick walls.
- Brick lintels are not structurally strong and they are used for small openings

generally not exceeding 1 meter span and light loads. They are built up with hand well burnt, copper colored, free from cracks and with sharp and straight edged bricks.

- Steel lintels are provided where the opening is large and the super imposed loads are also heavy. It consists of rolled steel joints or channel sections.
- Reinforced cement concrete lintels: Reinforced cement concrete lintels have replaced practically all other types of lintels because of their strength, rigidity, fire resistance economy and case in construction. Its width is kept equal to the width of the wall. The depth of R.C.C lintel and the reinforced depends upon the span and magnitude of loading. It can be pre-cast or cast-in-situ, pre-cast R.C.C lintels are preferred for small spans up to about 2 meters. The depth of lintel and the reinforcement is governed by the intensity of load, the type of support and the span.

1.8.4 Staircases

- Tread: The horizontal portion of a step was the foot rests is referred to as tread.250 to 300 mm is the typical dimensions of a tread.
- Riser: Riser is the vertical distance between the adjacent treads or the vertical projection of the step with value of 150 to 190 mm depending upon the type of building.
- Going: This is the horizontal projection of an inclined flight of steps between the first and last riser.

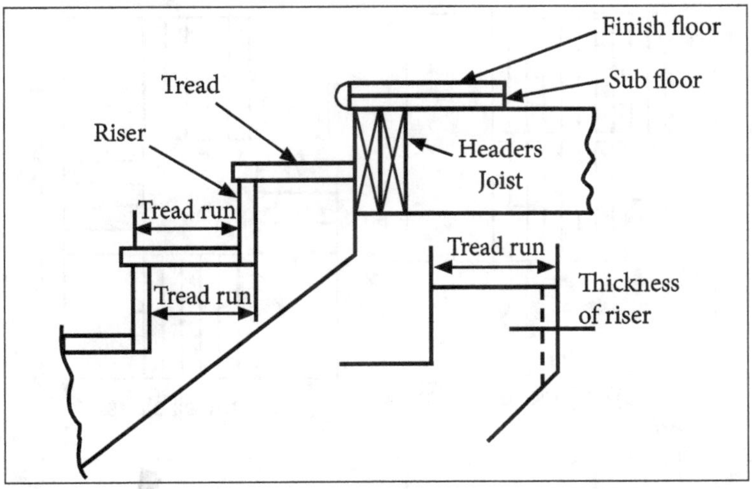

Types of Staircases

They are broadly classified as:

- Straight stair.
- Quarter turn stair.

- Half turn stair.
- Dog legged stair.
- Open newer stair with quarter space landing.
- Geometrical stairs such as circular stair, spiral stair, etc.

The below are the places where the following footings will be used:

- Single flight staircase: Single flight staircase is used in cellars or attics where the height between floors is small and the frequency of its use is less.
- Quarter turn staircase: Quarter turn staircase flight generally runs adjoining the walls and provides uninterrupted space at the centre of the room. Commonly used in domestic houses where floor heights are limited to 3m.

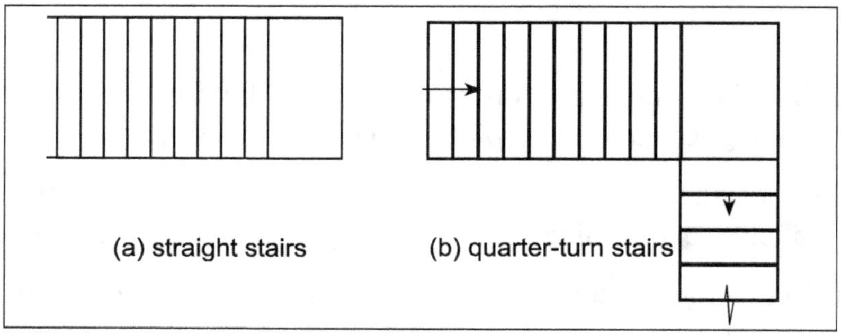

(a) straight stairs (b) quarter-turn stairs

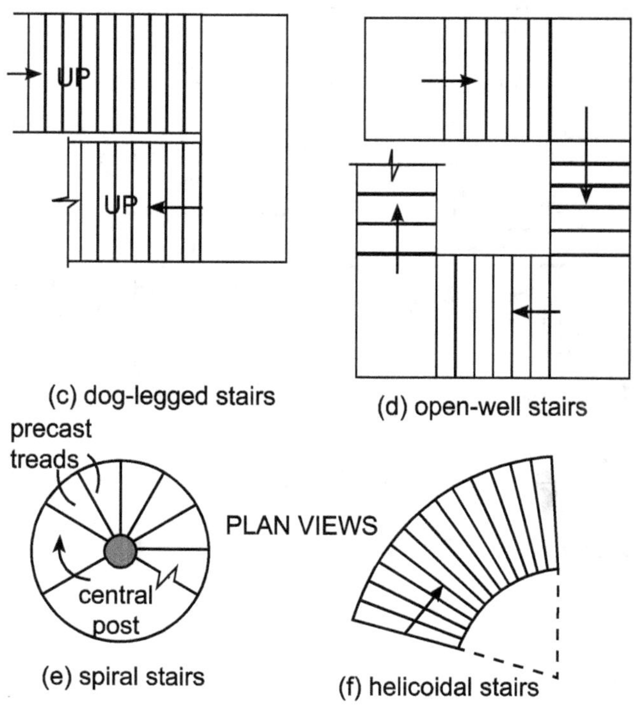

(c) dog-legged stairs (d) open-well stairs

(e) spiral stairs (f) helicoidal stairs

Common geometrical configurations of stairs

- Dog legged staircase: Dog legged staircase is usually adopted in economical utilization of available space.
- Open well staircase: Open well staircases are provided in public buildings where large spaces are available.
- Spiral staircase: In congested locations where space availability is small Spiral Stairs are provided.

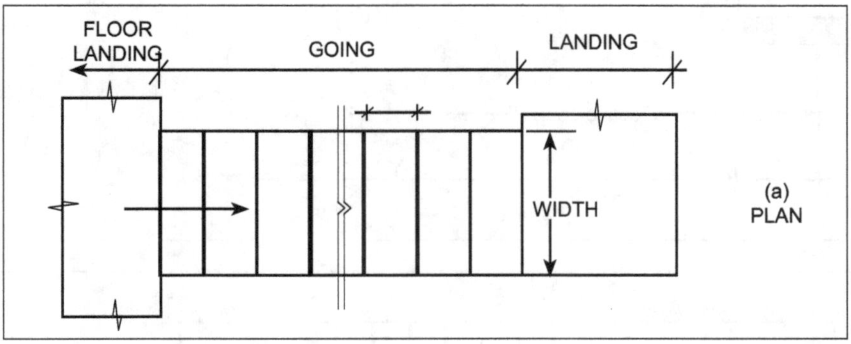

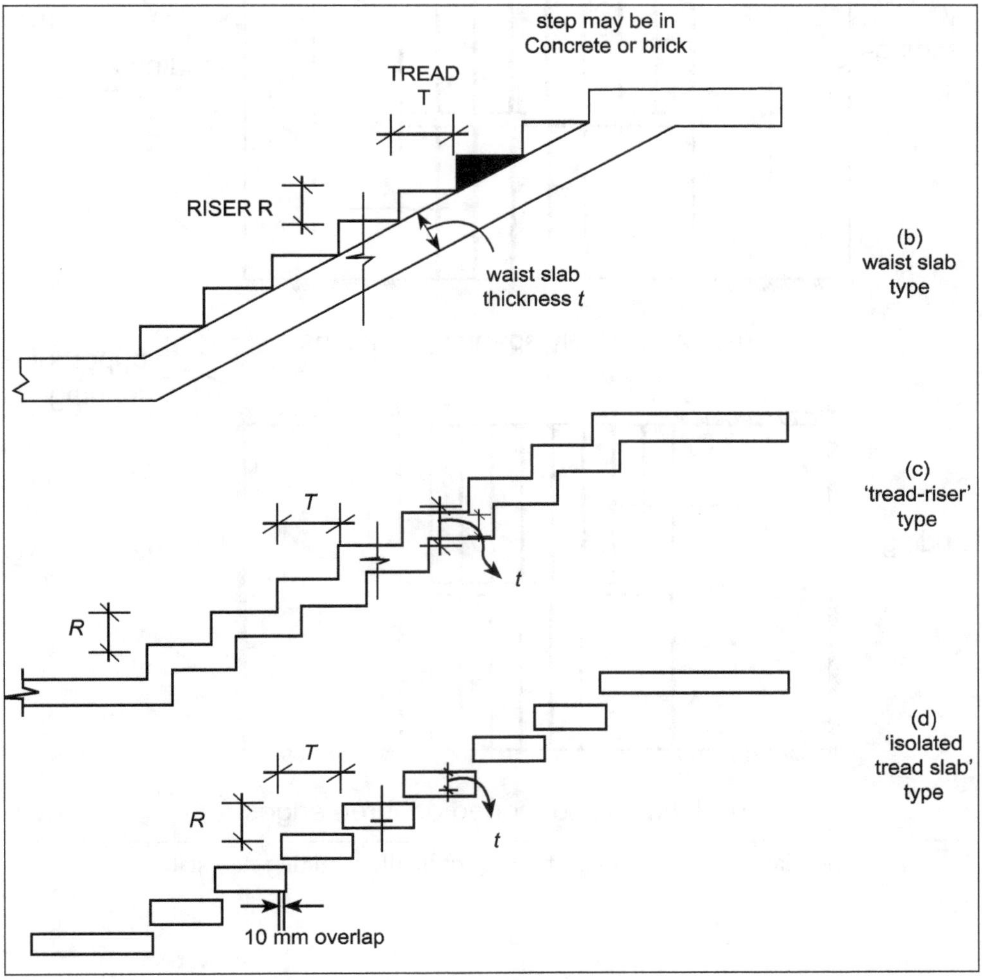

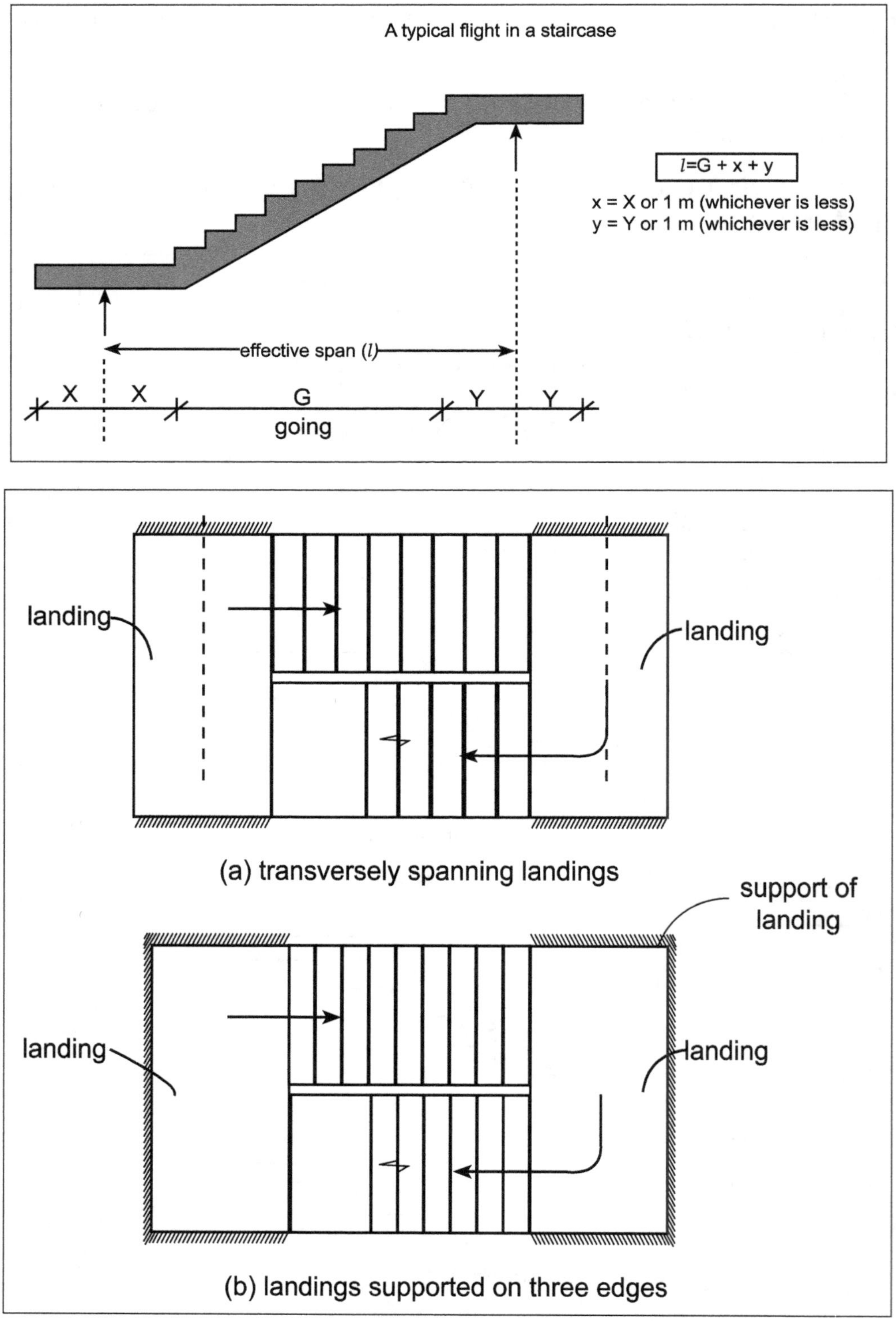

Special support conditions for longitudinally spanning stair slabs.

2

Surveying and Positioning

2.1 Introduction to Surveying Instruments

Surveying

Surveying is an art of determining the relative positions of various points on, above or below the surface of the earth by means of direct or indirect measurement of distance, direction and elevation.

Different methods and instruments are being used to facilitate the work of surveying.

The primary aims of field surveying are:

- To measure the horizontal distance between points.
- To measure the vertical elevation between points.
- To find out the relative direction of lines by measuring horizontal angles with reference to any arbitrary direction.
- To find out absolute direction by measuring horizontal angles with reference to a fixed direction.
- These parameters are utilized to find out the relative or absolute coordinates of a point / location.

Surveying.

Classifications

The various classifications of survey are explained below:

- Engineering survey: To collect requisite data for planning, design and execution of engineering projects. Three broad steps are:
    - Reconnaissance survey: To explore site conditions and availability of infrastructures.
    - Preliminary survey: To collect adequate data to prepare plan map of area to be used for planning and design.
    - Location survey: To set out work on the ground for actual construction execution of the project.
- Construction surveys: Surveys which are required for establishment of points, lines, grades and for staking out engineering works.
- Astronomic surveys: To determine the latitude, longitude and azimuth from astronomical observation.
- Mine surveys: To carry out surveying specific for open cast and underground mining.
- Chain survey: Chain survey is the simplest method of surveying. In this survey only measurements are taken in the field and rest of the work, such as plotting, calculation etc. are done in the office. This is most suitable and adapted to small plane areas with very few details. If carefully done, it gives quite accurate results. The necessary requirements for field work are chain, tape, ranging rod, arrows and sometime cross staff.
- Tie or subsidiary lines: A tie line joints two fixed points on the main survey lines. It helps to check the accuracy of surveying and to locate the interior details. The position of each tie line should be close to some features, such as paths, buildings etc.
- Base lines: It is main and longest line, which passes approximately through the centre of the field. All the other measurements to show the details of the work are taken with respect of this line.
- Check line: A check line also termed as a proof line is a line joining the apex of a triangle to some fixed points on any two sides of a triangle. A check line is measured to check the accuracy of the framework. The length of a check line, as measured on the ground should agree with its length on the plan.
- Offsets: These are the lateral measurements from the base line to fix the

positions of the different objects of the work with respect to base line. These are generally set at right angle offsets. It can also be drawn with the help of a tape.

- There are two kinds of offsets:
    - Perpendicular offsets.
    - Oblique offsets.

The measurements are taken at right angle to the survey line called perpendicular or right angled offsets. The measurements which are not made at right angles to the survey line are called oblique offsets or tie line offsets.

Procedure in Chain Survey

1. Reconnaissance

The preliminary inspection of the area to be surveyed is called reconnaissance. The surveyor inspects the area to be surveyed, survey or prepares index sketch or key plan.

2. Marking Station

- Surveyor fixes up the required number of stations at places, from where maximum possible stations are possible.
- Then he selects the way for passing the main line, which should be horizontal and clean as possible and should pass approximately through the centre of work.
- Then ranging roads are fixed on the stations.
- After fixing the stations, chaining could be started.
- Make ranging wherever necessary.
- Measure the change and offset.
- Enter in the field book.

Topographic Surveys

The purpose of a topographic survey is to gather survey data about the natural and man-made features of the land, as well as its elevations. From this information a three-dimensional map may be prepared. We may prepare the topographic map in the office after collecting the field data or prepare it right away in the field by plane table.

The work usually consists of the following:

- Establishing horizontal and vertical control that will serve as the framework of the survey.

- Determining enough horizontal location and elevation (usually called side shots) of ground points to provide enough data for plotting when the map is prepared.
- Locating natural and man-made features that may be required by the purpose of the survey.
- Computing distances, angles and elevations.
- Drawing the topographic map topographic surveys are commonly identified with horizontal and/or vertical control of third-and lower-order accuracies.

Route Surveys

The term route survey refers to surveys necessary for the location and construction of lines for transportation or communication that continues across country for some distance, such as highways, railroads, open-conduit systems, pipelines and power lines.

Generally, the preliminary survey for this work takes the form of a topographic survey. In the final stage, the work may consist of the following:

- Locating the center line, usually marked by stakes at 100-ft intervals called stations.
- Determining elevations along and across the center line for plotting profile and cross sections.
- Plotting the profile and cross sections and fixing the grades.
- Computing the volumes of earthwork and preparing a mass diagram.
- Staking out the extremities for cuts and fills.
- Determining drainage areas to be used in the design of ditches and culverts.
- Laying out structures, such as bridges and culverts.
- Locating right-of-way boundaries, as well as staking out fence lines, if necessary.

Special Surveys

Special surveys are conducted for a specific purpose and with a special type of surveying equipment and methods. A brief discussion of some of the special surveys familiar to us are as follows:

Land Surveys: Land surveys (sometimes called cadastral or property surveys) are conducted to establish the exact location, boundaries or subdivision of a tract of land in any specified area. This type of survey requires professional registration in all states.

Presently, land surveys generally consist of the following chores:

- Establishing markers or monuments to define and thereby preserve the boundaries of land belonging to a private concern, a corporation or the government.
- Relocating markers or monuments legally established by original surveys. This requires examining previous survey records and retracing what was done. When some markers or monuments are missing, they are re-established following recognized procedures, using whatever information is available.
- Re-running old land survey lines to determine their lengths and directions. As a result of the high cost of land, old lines are remeasured to get more precise measurements.
- Subdividing landed estates into parcels of predetermined sizes and shapes.
- Calculating areas, distances and directions and preparing the land map to portray the survey data so that it can be used as a permanent record.
- Writing a technical description for deeds.

Control Surveys: Control surveys provide "basic control" or horizontal and vertical positions of point to which supplementary surveys are adjusted. These types of surveys (sometimes termed as geodetic survey) are conducted to provide geographic positions and plane co-ordinates of triangulation and traverse stations and the elevations of benchmarks. These control points are further used as references for hydrographic surveys of the coastal waters, for topographic control and for the control of many state, city and private surveys. Horizontal and vertical controls generated by land surveys provide co-ordinated position data for all surveyors. It is therefore necessary that these types of surveys use first-order and second-order accuracies.

Hydrographic Surveys: Hydrographic surveys are made to acquire data required to chart and/or map shorelines and bottom depths of streams, rivers, lakes, reservoirs and other larger bodies of water.

This type of survey is also of general importance to navigation and to development of water resources for flood control, irrigation, electrical power and water supply.

2.1.1 Levels

Levelling may be defined as the art of determining the relative height or elevations of points or objects on the earth's surface.

Instruments used for levelling:

- Level.
- Levelling staff.

Level

The purpose of a level is to provide a horizontal line of sight.

Parts:

- A telescope to provide line of sight.
- A level tube to make the line of sight horizontal.
- A levelling head to bring the bubble in its centre of run.
- A tripod to support the instrument.

Dumpy Level

The telescope is rigidly fixed with the support and therefore can neither be rotated about the longitudinal axis nor can it be removed from its support. A long bubble tube is attached to the top of the telescope.

The levelling head generally consists of two parallel plates with either three foot screws or four foot screws. The upper plate is known as tribrach and the lower plate is known as trivet which can be screwed on a tripod.

Levelling Staff

A levelling staff is a straight rectangular rod having graduations, the foot of the staff representing zero reading.

Folding Type of 4m Levelling Staff

It consists of two wooden pieces each of 2m of length with the joint assembly. The thickness and width of staff are respectively 18m and 75mm. The folding joint is of the detachable type with locking device at the back. Hence the stand can be made into two parts for easy handling. When two pieces are locked together the two pieces form a straight rigid levelling staff of length equal to 4m.

Principle of Levelling

When the level is set up correctly and levelled, the line of collimation will be horizontal. Telescope is rotated about its vertical axis, it will revolve in a horizontal plane known as the plane of collimation and therefore all staff readings taken will be the vertical measurements, made downwards from this plane.

To find by how much amount the line of sight is above the bench mark and to ascertain by how much amount the next point is below or above the line sight.

Height of instrument= Elevation of B.M + Back sight.

Elevation of point B = Height of instrument – Foresight.

Reduction Level

Height of Instrument Method: In this method the height of instrument is calculated for each setting of instrument and then the elevation or reduced level of the turning point is calculated with respect to the height of the instrument.

Rise and Fall Method: In rise and fall method the difference of level between consequence point is found comparing staff readings on the two points of same setting of the instrument.

Plane Table Survey

Plane table survey is that type of survey in which the measurement of survey lines of the transverse and their plotting to a suitable scale are done simultaneously on the field. It consists of the following.

- Drawing board mounted on a tripod.
- Alidade.

Drawing Board

The drawing board is made of well-seasoned wood such as teak and its size varies from 400mm x 300mm to 750mm x 600 mm. The board is mounted on a tripod with ball and socket arrangement which allow the board to be levelled and rotated about the vertical axis.

Alidades

- The open slight alidade.
- The telescope alidade.

The open slight alidade consists of a flat rectangular piece of wood of about 25mm width, 12.5mm thickness and length varying from 200 to 500 mm. The levelled edge of the alidade is called fiducial edge. Its edge is graduated and serves as a scale. Vertical slight vanes are mounted at both ends. One of the slight vanes is proved with a narrow slit and the other with a central vertical wire. The line of sight is in the same vertical plane as the fiducial edge.

The telescope alidade consists of a telescope mounted on a horizontal axis, the ends of which are supported on standards. A circular level or two spirit levels are attached to the base of the telescope to level the plane table.

Working operations:

- Fixing: Fixing the table to the tripod stand.
- Setting: The table is set up at a convenient height say 1m above ground. The legs of the stand are spread apart and firmly fixed into the ground. The table is then centered. This means that the point plotted on the sheet corresponding to the station occupied should be exactly above the station on the ground. This is done by means of the plumbing fork.
- Leveling: The table is then levelled either by ordinarily tilting the board or by ball and socket arrangement.
- Orientation: Orientation is the process of putting the plane table into some fixed direction so that line representing a certain direction on the drawing sheet is parallel to the direction on the ground.

  Two methods adopted:

  - Orientation by magnetic needle.
  - Orientation by back sighting.

Methods of plane table survey:

- Radiation.
- Intersection.
- Traversing.
- Resection.

2.1.2 Theodolites

Theodolite is used to measure the horizontal and vertical angles. Theodolite is more precise than magnetic compass.

Applications

Theodolites are mainly used for surveying, but they are also useful in these applications:

- Navigating.

- Meteorology.
- Laying out building corners and lines.
- Measuring and laying out angles and straight lines.
- Aligning wooden frame walls.
- Forming panels.
- Plumbing a column or building corner.

Types

There are different types of theodolite available. It may be classified into three broad categories. They are:

- Vernier or transit theodolite.
- Digital theodolite.
- Total station.

Each type of theodolite is peculiar in its construction and mode of operation. However, inherent fundamentals of all are same. The vernier type theodolite which is most popular and is being widely used. The salient parts of a vernier theodolite have been discussed below:

- Leveling head.
- Shifting head.
- Lower plate.
- Upper plate.
- Plate levels.
- Standard (or a frame).
- Vernier frame.
- Telescope.
- Vertical circle.
- Altitude bubble.
- Screws.
- Tripod stand.

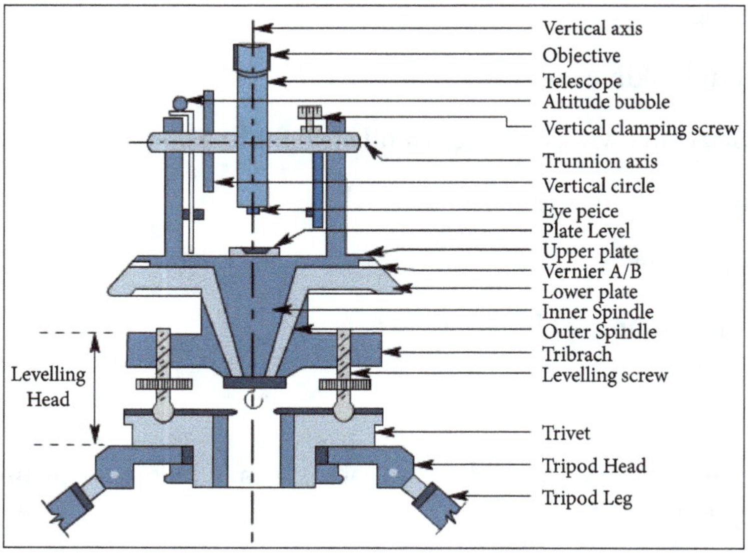

Sectional view of theodolite.

Vernier Transit Theodolite

A vernier transit theodolite has a telescope that flips over to allow back sighting and angle doubling, which is believed to result in fewer errors in reading. However, vernier transit theodolites are considered to be less precise than other types because they do not have features like scale magnification or measurements in micrometers.

Vernier transits are commonly used on construction sites because they are relatively lightweight and easily moved around. Although there are some vernier transit theodolites which measure both horizontal and vertical angles, some only measure horizontally.

Digital Theodolite

Digital theodolite.

This type of theodolite provides the value of observation directly in viewing panel. The precision of this type of instrument varies in the order of 1" to 10".

2.1.3 Plane Tables and Related Devices

The principle of plane tabling is parallelism, meaning that the rays drawn from stations to objects on the paper are parallel to the lines from the stations to the objects on the ground. The relative positions of the objects on the ground are represented by their plotted positions on the paper and lie on the respective rays. The table is always placed at each of the successive stations parallel to the position it occupied at the starting station. Plane tabling is a graphical method of surveying there the field work and plotting are done simultaneously and such survey does not involve the use of a field book.

Plane table survey is mainly suitable for filling interior details when traversing is done by theodolite, sometimes traversing by plane table may also be done. But this survey is recommended for the work where great accuracy is not required. As the fitting and fixing arrangement of this instrument is not perfect, most accurate work cannot be expected.

The following accessories are required to carry out plane table survey:

1. Alidade

It is a straight edge ruler having some form of sighting device. One edge of the ruler is leveled and is graduated. Always this edge is used for drawing line of sight. Depending on the type of line of sight there are two types of alidade:

- Plain Alidade: Figure shows a typical plain alidade. A sight vane is provided at each end of the ruler. The vane with narrow slit serves as eye vane and the other with wide slit and having a thin wire at its centre serves as object vane. The two vanes are provided with hinges at the ends of ruler so that when not in use they can be folded on the ruler. Plain alidade is not suitable in surveying hilly areas as the inclination of line of sight in this case is limited.

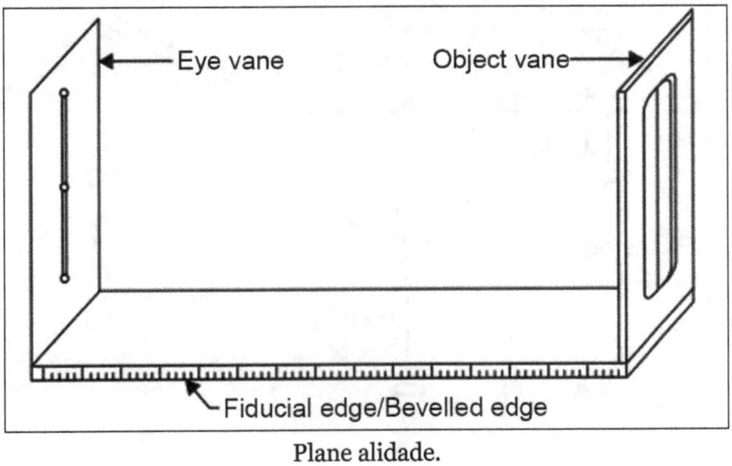

Plane alidade.

- Telescopic alidade: It consists of a telescope mounted on a column fixed to the ruler. The line of sight through the telescope is kept parallel to the bevelled edge of the ruler. The telescope is provided with a level tube and vertical graduation arc. If horizontal sight is required bubble in the level tube is kept at the centre. If inclined sights are required vertical graduation helps in noting the inclination of the line of sight. By providing telescope, the range and the accuracy of line of sight is increased.

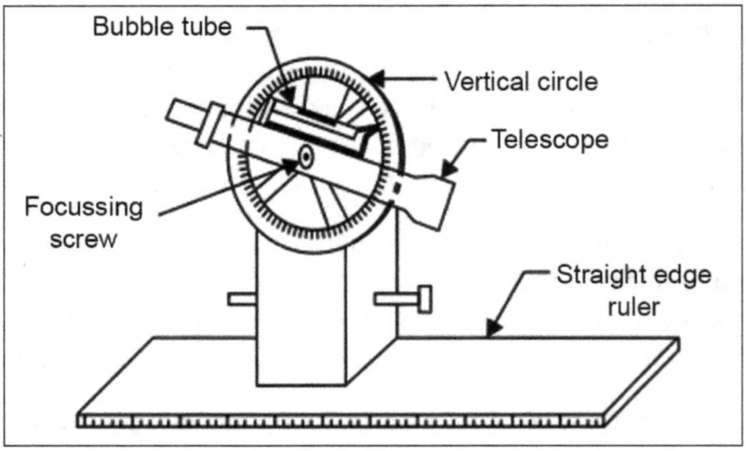

Telescopic alidade.

2. Plumbing Fork and Plumb Bob

Figure shows a typical plumbing fork with a plum bob. Plumbing fork is a U-shaped metal frame with a upper horizontal arm and a lower inclined arm. The upper arm is provided with a pointer at the end while the lower arm is provided with a hook to suspend plumb bob. When the plumbing fork is kept on the plane table the vertical line (line of plumb bob) passes through the pointed edge of upper arm. The plumb bob helps in transferring the ground point to the drawing sheet and vice versa also.

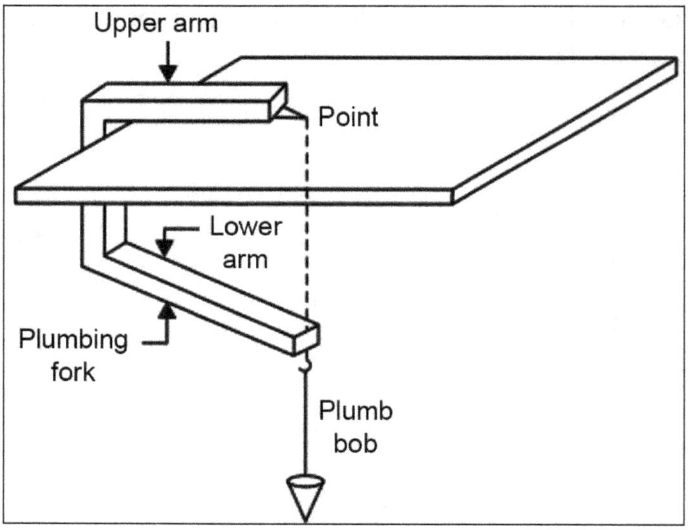

Plumbing fork and plumb bob.

3. Spirit Level

A flat based spirit level is used to level the plane table during surveying. To get perfect level, spirit level should show central position for bubble tube when checked with its positions in any two mutually perpendicular directions.

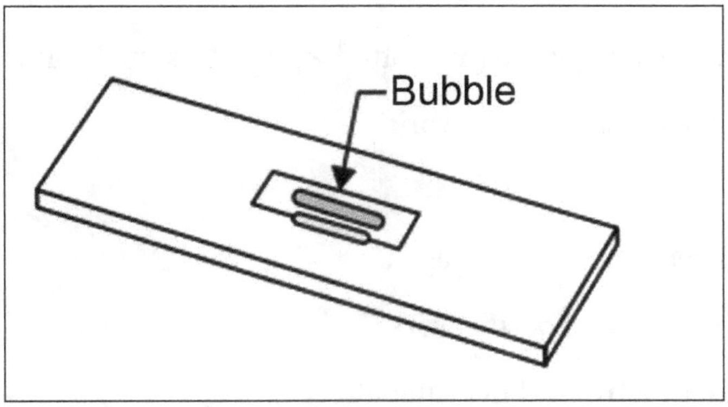

Spirit level.

4. Trough Compass

It consists of a 80 to 150 mm long and 30 mm wide box carrying a freely suspended needle at its centre. At the ends of the needle, graduations are marked on the box to indicate zero to five degrees on either side of the centre. The box is provided with glass top to prevent oscillation of the needle by wind. When needle is centered (reading 0–0), the line of needle is parallel to the edge of the box. Hence marking on the edges in this state indicates magnetic north–south direction.

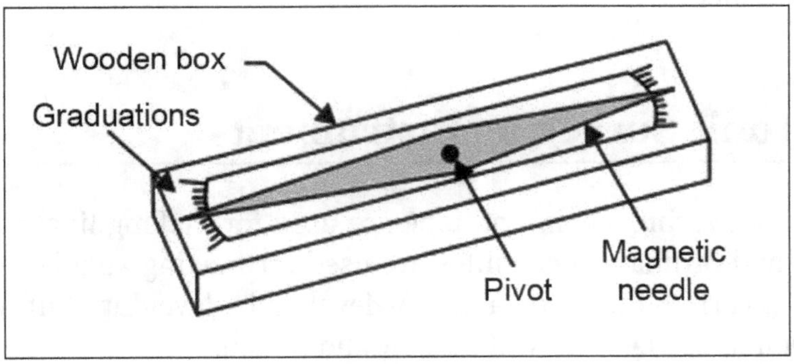

Trough compass.

5. Drawing Sheet and Accessories for Drawing

A good quality, seasoned drawing sheet should be used for plane table surveying. The drawing sheet may be rolled when not in use, but should never be folded. For important works fiber glass sheets or paper backed with thin aluminum sheets are used.

Clips clamps, adhesive tapes may be used for fixing drawing sheet to the plane table. Sharp hard pencil, good quality eraser, pencil cutter and sand paper to keep pencil point sharp are other accessories required for the drawing work. If necessary, plastic sheet should be carried to cover the drawing sheet from rain and dust.

Advantages:

- The plan is drawn at the field itself.
- The surveyor can compare the plotted work with actual features of the area.
- Most suitable for small scale work.

Disadvantages:

- Most inconvenient in rainy season.
- It is not intended for specific work.
- It is essentially a tropical instrument.

Methods of Plane Table Surveying

Four classes of plane tabling surveys are recognized as follows:

- Radiation method.
- Intersection method.
- Traversing method.
- Resection method.

2.2 Electronic Surveying Instruments

In conventional surveying, chain and tape are used for making linear measurements while compass and ordinary theodolites are used for making angular measurements. Levelling work is carried out using a dumpy level and a levelling staff. With such surveying instruments, survey work will be slow and tedious.

Hence modern surveying instruments are becoming more popular and they are gradually replacing old surveying instruments such as compass and dumpy level. With modern surveying instruments, survey work will be precise, faster and less tedious. Some of the modern surveying instruments are discussed in brief.

Electronic Distance Measurement (EDM) Instruments

Direct measurement of distances and their directions can be obtained by using electronic instruments that rely on propagation, reflection and reception of either light waves or radio waves. They may be broadly classified into three types:

1. Infrared Wave Instruments

These instruments measure distances by using amplitude modulated infrared waves. At the end of the line, prisms mounted on target are used to reflect the waves. These instruments are light and economical and can be mounted on theodolites for angular measurements. The range of such an instrument will be 3 km and the accuracy achieved is ± 10 mm. Eg. DISTOMAT DI 1000 and DISTOMAT DI 5.

- DISTOMAT DI 1000: It is a very small, compact EDM, particularly useful in building construction and other civil engineering works, where distance measurements are less than 500 m. It is an EDM that makes the meaning tape redundant. To measure the distance, one has to simply point the instrument to the reflector, touch a key and read the result.

2. Light Wave Instruments

These are the instruments which measures distances based on propagation of modulated light waves. The accuracy of such an instrument varies from 0.5 to 5 mm / km distance and has a range of nearly 3 km. For example:

- Geodimeter: The geodimeter is often used for night time observations and can measure distances up to 3 kilometers. As this instrument measures distances based on the propagation of modulated light waves, it requires a prism system at the end of the distance we are measuring to reflect the waves.

3. Microwave Instruments

These instruments make use of high frequency radio waves. These instruments were invented as early as 1950 in South Africa by Dr. T.L. Wadley. The range of these instruments is up to 100 km and can be used both during day and night. For example:

- Tellurometer: It is an EDM which uses high frequency radio waves (micro-waves) for measuring distances. It is a highly portable instrument and can be worked with 12 to 24 volt battery. For measuring distance, two tellurometers are required, one to be stationed at each end of the line, with two highly skilled persons, to take observations. One instrument is used as a master unit and the other as a remote unit. Just by pressing a button a master can be converted in to remote unit and vice-versa. A speech facility (communication facility) is provided to each operator to interact during measurement.

Total Station

It is a light weight, compact and fully integrated electronic instrument combining the capability of an EDM and an angular measuring instrument such as wild theodolite. It can perform the following functions:

- Distance measurement.
- Angular measurement.
- Data processing.
- Digital display of point details.
- Storing data is an electronic field book.

The important features of total station are as follows:

- Key-board-control: All the functions are controlled by operating key board.
- Digital panel: The panel displays the values of distance, angle, height and the co-ordinates of the observed point, where the reflector (target) is kept.
- Remote height object: The heights of some inaccessible objects such as towers can be read directly. The micro-processor provided in the instrument applies the correction for earth's curvature and mean refraction, automatically.
- Traversing program: The co-ordinates of the reflector and the angle or bearing on the reflector can be stored and can be recalled for next set up of instrument.
- Setting out for distance, direction and height: Whenever a particular direction and horizontal distance is to be entered for the purpose of locating the point on the ground using a target, then the instrument displays the angle through which the theodolite has to be turned and the distance by which the reflector should move.
- Global Positioning System: This system is developed by U.S. Defense department and is called Navigational System with Time and Ranging Global Positioning System (NAVSTAR GPS) or simply GPS. For this purpose U.S. Air Force has stationed 24 satellites at an attitude of 20,200 km above the earth's surface. The satellites have been positioned in such a way; at least four satellites will be visible from any point on earth. The user needs a GPS receiver to locate the position of any point on ground. The receive processes the signals received from the satellite and compute the position (latitude and longitude) and elevation of a point with reference to datum.
- Automatic level: An automatic level is a special surveying (levelling) instrument which contains an optical compensator which maintains line of sight or line of collimation even though instrument is slightly tilled. Eg: Wild NAK2 Automatic level.

2.3 Measurement of Distances: Conventional and EDM Methods

This deals with the principles and procedures of separate electronic distance measuring device (EDM) instruments that are mountable with optic/electronic theodolites. Original EDM's were individual units used only for distance measurements. EDM instruments used as a modular component in modern total stations also work under the same principle of separate EDM instruments.

EDM was first introduced in the 1950's and has undergone several modifications. The early versions of the instruments, though very precise over long distances, were very large, heavy, complicated and expensive. Electronic theodolite/optic theodolite mountable EDM was introduced first, and these are being used still today. Rapid advances in related technologies have provided lighter, simpler and less expensive instruments. Latest versions of EDM instruments are used as a modular component of total stations. Some earlier versions of EDM instruments used natural light for the calculation of distances, but the latest version of EDM uses infrared light, laser light or microwaves.

Measurement Principle of EDM Instrument

The principle of the measurement device in EDM, which is currently used in a total station and used along with electronic/optic theodolites, is that it calculates the distance by measuring the phase shift during the radiated electromagnetic wave (such as an infrared light or laser light or microwave) from the EDM's main unit, which returns by being reflected through the reflector, which is positioned at a measurement point. This phase shift can be regarded as a part of the frequency that appears as the unit of time or length under a specific condition.

When the slope distance L and the slope angle ϕ are measured by EDM, if the elevation of point A is the reference point, we can find the elevation of point B by the following formula (equation 1).

$$\text{Elevation of Point B} = \text{Elevation of Point } A + HI \pm L \sin \varphi - HR \qquad ...(1)$$

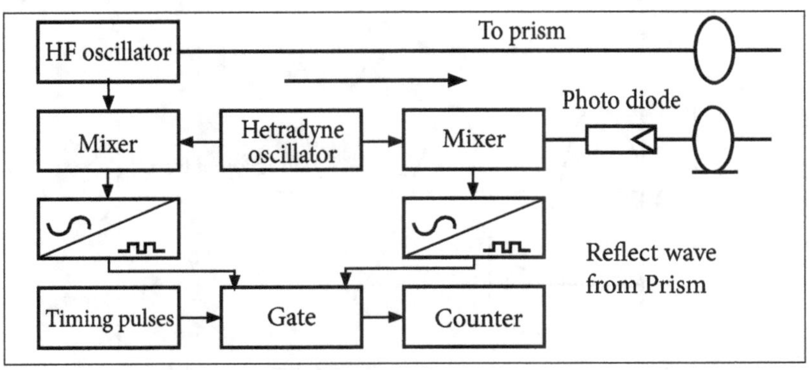

EDM's structure.

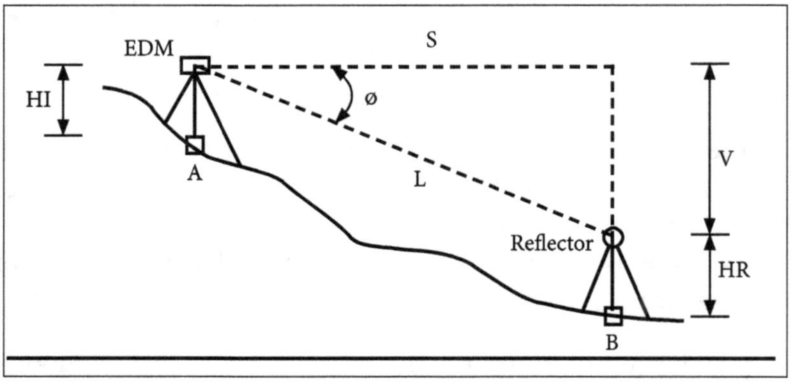

Slope distance (horizontal distance error) by incidence angle.

EDM wave shows a wave of wavelength λ. The wave is travelling along the X axis with a velocity of 299,792,560.4 km/sec (approximate velocity of light in vacuum). The frequency of the wave is the time taken for one complete wavelength.

$$\lambda = \frac{c}{f}$$

Where,

λ = Wavelength in meters.

c = Velocity in km/sec.

f = Frequency in hertz (one cycle per second).

Principle of electronic distance measurement shows a modulated electromagnetic wave being emitted from an EDM instrument and being reflected back to the instrument. Here the double distance is taken as 2L, which is equal to the total whole number of wavelength nλ and the partial wavelength ω. Therefore, the distance between the EDM instrument and the reflector (L) is calculated as follows:

$$L = \frac{n\lambda + \varphi}{2} \text{ meters}$$

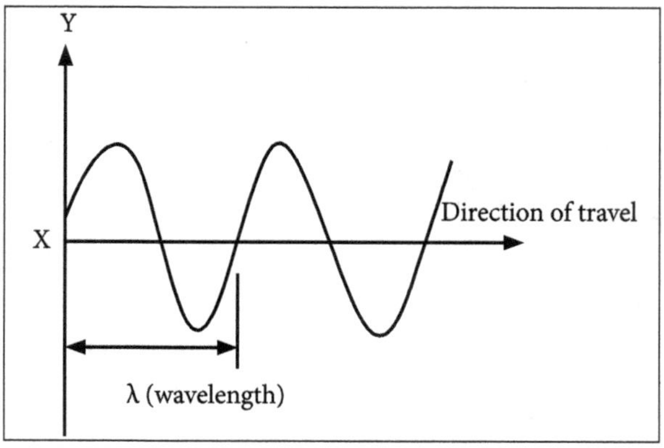

EDM wave.

The partial wavelength (ω) is determined by calculating the phase shift required by correlating the transmitted and reflected waves, that is, by calculating the phase delay required to match precisely the transmitted and the reflected waves:

S = Station

r = Reflector component of addition constant

Z = Target

E = References plane within the EDM for phase comparison

λ = Modulation of wavelength

ω = Fraction to be measured of a whole wavelength of modulation Δ λ

e = Distance meter component of addition constant.

R = Reference plane for the reflection of the wave transmitted by the EDM.

The instrument transmits a series of three or four modulated waves at different frequencies. By substituting the resulting values of λ and ω in the above equation for three or four different frequencies, the value of n can be found. The instruments are designed to carry out this procedure within seconds, and display the value of L. Some EDM instruments use pulsed laser emissions, and these instruments determine the distance by measuring the time taken between the transmission of the signal and the reception of the reflected signal, by making use of the pulsed laser beam.

The velocity of light (electromagnetic waves) through the atmosphere can be affected by the following:

- Temperature.
- Atmospheric pressure.
- Water vapour content in the atmosphere.

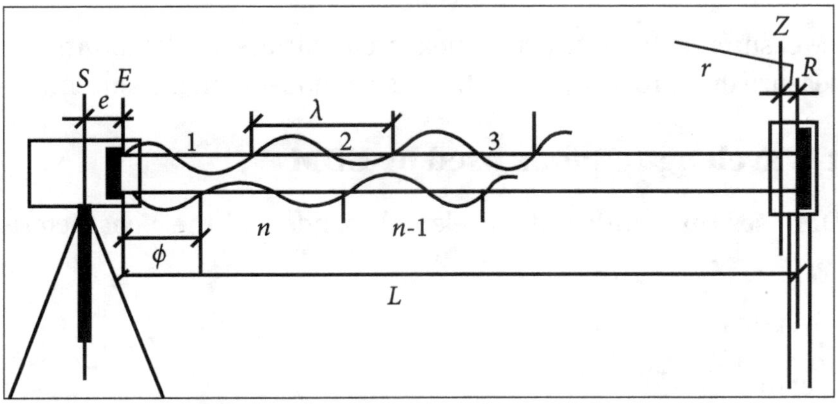

Principle of electronic distance measurement.

Note: The components of e and r are only auxiliary quantities.

The corrections for temperature and pressure are determined manually by referring the monograms given with all total stations to electronic distance measurement, or the corrections are calculated automatically in the instrument itself by inputting the values for temperature and pressure.

For measuring short distances using EDM instruments, atmospheric corrections are relatively insignificant. For measuring long distances, atmospheric corrections become quite important.

EDM Instrument Characteristics

The characteristics of EDM instruments are as mentioned below:

Distance Range: Distance can be measured up to 1 km using a single prism under average atmospheric conditions. Short range EDM instruments can measure up to 1250 m using a single prism. Long range EDM instruments can be used for the measurement up to 15 km using prisms.

Accuracy:

- For short range EDM instruments: ±15 mm + 5 ppm.
- For long range EDM instruments: ±3 mm + 1 ppm.

Measuring Time: The measuring time required is 1.5 sec for short range measurements and up to 4 sec for long range measurements. Both accuracy and time are considerably reduced for tracking mode measurements.

Slope Reduction: Manual or automatic in some models. The average of repeated measurements is available on some models.

Battery Capability: 1500–5000 measurements are depending on the power of the battery and the temperature.

Non-prism Measurements: Non-prism measurements are available with some models. They can measure up to 100–350 m in the case of non-prism measurements.

Different Wavelength Bands used by EDM

Usually, EDM uses three different wavelength bands and their characteristics are as follows.

Microwave Systems:

- Range up to 150 km.

- Wavelength 3 cm.
- Not limited to line of sight.
- Unaffected by visibility.

Light wave systems:

- Range up to 5 km (for small machines).
- Visible light, lasers.
- Distance reduced by visibility.

Infrared systems:

- Range up to 3 km.
- Limited to line of sight.
- Limited by rain, fog and other airborne particles.

Prisms used for Electronic Distance Measurements

Prisms are used to reflect transmitted signals from an EDM instrument. A single reflector is a cube corner prism that has the characteristics of reflecting light rays precisely back to the emitting EDM instrument. This retro-direct capability means that the prism can be somewhat misaligned with respect to the EDM instrument and still be effective. Cutting the corners off a solid glass cube forms a cube corner prism. The quality of the prism is determined by the flatness of the surface and the perpendicularity of the 90° surfaces. Prisms can be tribrach-mounted on a tripod, centered by an optical plummet, or attached to a prism pole held vertical on a point with the aid of a bull's eye level. If higher accuracy is required, then the prisms must be tribrach-mounted. For control surveys, tribrach of prisms can be detached and then interchanged with theodolite (and EDM), similarly mounted at the other end of the line being measured. This interchangeability of prism and total station speeds up the work, as the tribrach mounted on the tripod is centered and levelled only once. Equipment that can be interchanged and mounted on a tribrach already set up is known as forced centering equipment.

A high precision single prism target.

A mini prism.

A single prism mounted on a tribrach.

Different types of prism reflectors.

Prisms mounted on adjustable-length prism poles are portable and are well suited for stakeout surveys and topographic surveys. It is very important that the prisms mounted on poles or tribrach be permitted to tilt up and down, so that they can be made perpendicular to EDM signals sent from much higher or lower positions. This characteristic is particularly important for short sights.

Accuracy in Electronic Distance Measurements

EDM instrument accuracies are stated in terms of an instrumental error and a measuring error, which are proportional to the distance measured. Typically, accuracy is given as ± {5 mm + 5 ppm (parts per million)}. Here, the first numerical ±5 mm is the instrument error, which is independent of the measurement, and the second numerical

5-ppm (i.e. 5 ppm = 5 mm/km) denotes the distance related error. High end receivers are getting accuracies up to ± (3 mm + 1 ppm), whereas, the low end receivers are getting accuracies up to ± (10 mm + 10 ppm). The second part of the error, known as a proportional part error (10ppm), is insignificant for most works, and the constant part of the error ± (5 mm) is less significant as the distance being measured lengthens. At 100 m, an error of ±5 mm represents 1/20,000 accuracy, whereas, for 1000 m, the same instrument error represents 1/200,000 accuracy.

For most accurate works, both the EDM instrument and the prism reflectors must be corrected for off-center characteristics. The measurement is recorded from the electrical center of the EDM instrument to the back of the prism and back to the electrical center of the EDM instrument. The manufacturer compensates for the difference between the electrical center of the EDM instrument and the plumb line through the tribrach center at the factory itself. The prism constant is eliminated either by the EDM instrument manufacturer at the factory or in the field.

Field Procedure of Electronic Distance Measurement

The field procedure of electronic distance measurement includes the following:

The setup of the EDM instruments are fixed on a tribrach (for tribrach-mounted EDM instruments), after the tribrach has been set over a required point on the surface of the earth. Centering should be carried out using an optical plummet. Theodolite yoke-mounted EDM instruments are simply attached to the theodolite before or after the theodolite has been set over a required station point. Prisms are set over the remote station points by either inserting the prism to a tribrach after centering, or by holding the prism vertically over the required point with the help of a prism pole. The EDM instrument is then turned on. A quick check for battery, display etc. is made to ensure that it is functioning properly. The height of the instrument and the height of the prism center are measured and recorded.

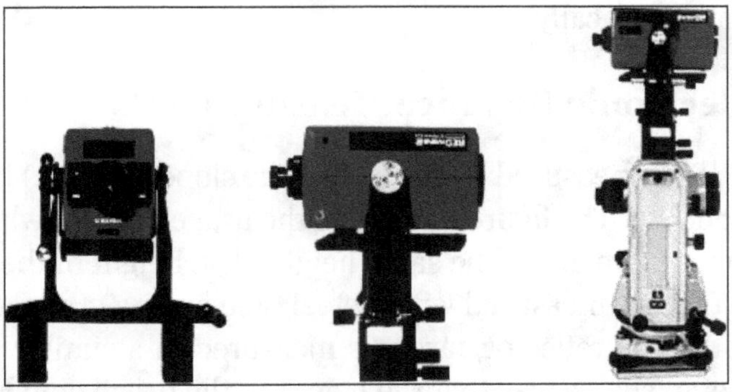

Two side views of an EDM with fixing clamps are fixed in an electronic theodolite.

Sighting the Target and Aiming

The target is sighted by orienting the EDM instrument (orienting the telescope in the case of theodolite yoke-mounted EDM instrument). Telescope or yoke-mounted EDM instruments have the optical line of sight a bit below than the electronic signal. The surveyor can set the electronic signal precisely on the prism center by adjusting the appropriate horizontal and vertical slow motion screws. This can be done with respect to the audible maximum signal intensity level from the EDM instrument, or according to the graphical display of the EDM instrument. Older versions of EDM instruments are provided with an attenuator that has to be adjusted for varying distances, in which the strength of the signal has to be reduced for short distances and increased for long distances to avoid signal overload.

Measurement

Before taking measurements, the vertical angle has to be fed into the EDM instrument. Slope distance measurement is accomplished by pressing the measure button of the EDM instrument. Within a couple of seconds, the result will appear on the display of the EDM instrument. Usually, the EDM instruments have a liquid crystal display (LCD). If no measurements appear in the display, then the surveyor should check the switch position, battery status, attenuation and target sight. EDM instruments with a built-in microprocessor will show the horizontal and vertical distances, coordinates etc. The input data like vertical angle, ppm-correction and prism constant are entered through the keyboard of the EDM instrument. Most of the EDM instruments have a tracking mode, which helps in the case of layout surveys and permits continuous distance update as the prism is moved closer to its final layout position. Surveyors may use hand-held radios for communication for long distance measurements.

Record

The measured data can be recorded conveniently in field note format, or they can be manually entered into an electronic data collector. The distance data must be accompanied by all relevant atmospheric and instrumental correction factors. Total station instruments, which have automatic data acquisition capabilities, will record and store the measurement automatically.

Geometry of Electronic Distance Measurements

The slope distance (S) is measured by the EDM. The slope angle (α) is measured by the accompanying theodolite. The figure illustrates the usage of EDM when the optical target and the reflecting prism are at the same height. The height of the EDM instrument and the theodolite (hi) are measured with a steel tape or by a graduated tripod-centering rod. The height of the reflector/target is measured in a similar way. If the height of the reflector is set in such a way that the height of the prism equals the height of the instrument (hi), it will simplify the computations. Nothing will happen even if the target height and the instrument height are unequal. The elevation of the station is known and the elevation of station B is calculated as follows:

$$\text{Elevation of station B} = \text{Elevation of station A} + hi \pm V - hr$$

When the EDM instrument is mounted on the theodolite and the target is located below the prism, the geometric relationship. The additional problem encountered in this situation is the computation of the correction of the vertical angle ($\Delta\alpha$) that occurs when Δhi and Δhr are different. The precise size of the vertical angle is important because it is used in conjunction with the measured slope distance to compute the horizontal and vertical distances. The difference between Δhr and Δhi is '$X = \Delta hr - \Delta hi$'. The small triangle formed by extending S' has the hypotenuse equal to X and an angle of $\Delta\alpha$.This permits computation of the side $X \cos\alpha$, which can be used together with S to determine $\Delta\alpha$.

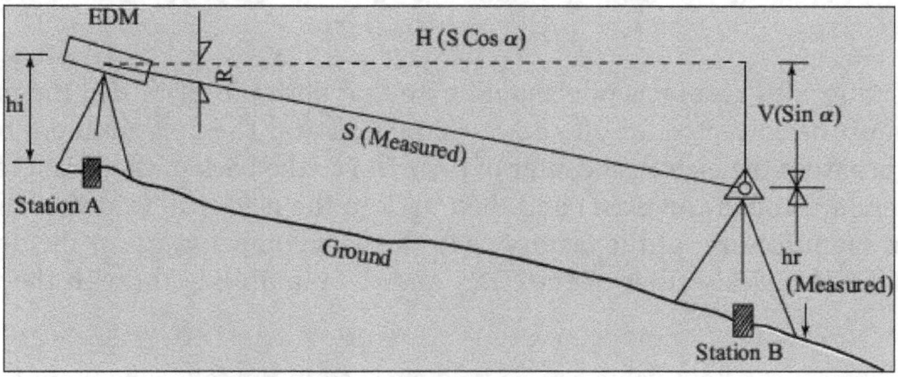

Geometry of an EDM measurement.

$$\frac{X\cos\alpha}{S} = \sin\Delta\alpha$$

Focusing and Sighting

The reticule cross hairs are focused by pointing the telescope at some other uniformly light surface (the sky), and by rotating the inner black portion of the eyepiece until the cross hairs are sharp and black.

The telescope is pointed towards the reflector by means of optical sight. Use of this sight will speed up acquisition of the reflector significantly. With both eyes open, one can have the image of the white cross hair in the optical sight in one eye, and the reflector in the other. When the two are superimposed, one is approximately on the target. This saves a significant amount of time searching for the reflector with the actual telescope.

The horizontal and vertical rotations of the theodolite are controlled in two ways. First, we can rotate the theodolite horizontally by hand by placing two hands on either side of it and rotating it. To rotate vertically, the EDM unit is rotated up or down by hand by nudging it up or down.

EDM Accuracies

EDM accuracies are stated in terms of constant instrumental errors and a measuring error proportional to the distance being measured.

Typically, accuracy is claimed as (65 mm + 5 ppm). The 65 mm is the instrumental error that is independent of the length of the measurement, whereas the 5 ppm (5 mm/km) denotes the distance related error.

Most of the instruments now available are claiming accuracies in the range of 6(3 mm + 1 ppm) to 6(10 mm +10 ppm). The pre-operational part error (ppm) is insignificant in most of the work, and the constant part of the error is less significant when the distances being measured are more. At 100 m, an error of 65 mm represents 1/20,000 accuracy, whereas at 1,000 m, the same instrumental error represents 1/200,000 accuracy.

When dealing with accuracy, one should note that both the EDM and the prism reflectors must be corrected for off-center characteristics. The measurement being recorded goes from the electrical center of the EDM to the back of the prism (allowing for refraction through the glass) and then back to the electrical center of the EDM. The EDM manufacturer at the factory or in the field compensates for the difference between the electrical center of the EDM and the plumb line through the tribrach center.

The EDM/prism constant value can be field checked in the following manner. A long line (more than 1 km length) is laid out with end stations and an intermediate station.

The overall distance AC is measured, along with partial lengths AB and BC. The constant value will be present in all measurements. Therefore:

AC – AB – BC = Instrument/prism constant.

Alternatively, the constant can be determined by measuring a known baseline, if one can be conveniently accessed.

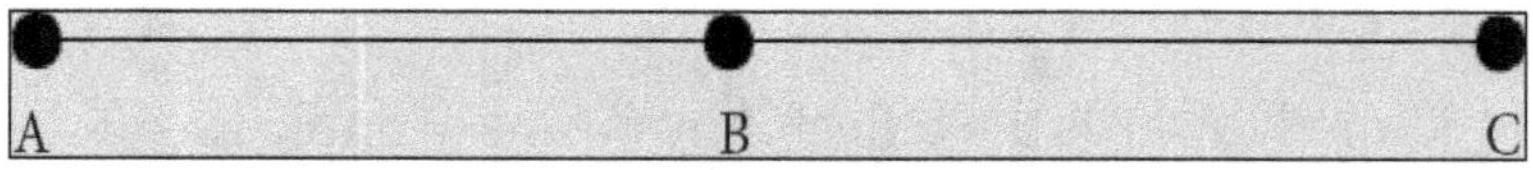

Method of determining the instrument-reflector constant.

2.4 Measurement of Directions and Elevations by Different Methods

1. Horizontal Distance

Tachymetry a rapid optical means of measuring distance using a telescope with cross hairs and a stadia rod (one stadium = about 607 feet).

Measuring of slope distances with a tape and reducing it to horizontal distance using the cosine of the slope gradient.

2. Difference In Elevation

Levelling with a level telescope and a stadia rod, or measuring a vertical angles and a slope distance (height is the product of the distance and sine of the angle).

Levelling is more accurate since elevation differences are measured and not calculated. Two readings are taken at each position of the automatic level: a back sight towards a station located before the level on the traverse and a foresight to the next station on the traverse. Thus the stadia rod occupies two stations, before and after the level on the survey. The difference in elevation between successive stations if the difference between the back sight and the foresight read from the stadia rod. For each position of the level, the lengths of the foresights and back sights should be approximately the same since accuracy is a function of the distance of a sighting. The level operator should anticipate the distance to the next station and set up the level midway along this distance (note: the distance that can be sighted decreases with increase in slope, since the stadia rod will disappear above or below the level line of sight). The lengths of back sight and foresight can be paced by the rod person or measured by the interval between the upper and lower cross hairs (tachymetry).

3. Direction

- Horizontal angle measured with a compass.
- Precise measuring devices use vernier scales.
- Direction is expressed relative to a reference line or meridian:
    - True meridian: A north-south line.
    - Magnetic meridian: A line parallel with the earth's magnetic lines of force.
    - Assumed meridian: An arbitrary line.

Types of Horizontal Angles

- Bearing: Angles expressed relative to a meridian using the quadrant and an acute angle, e.g. N37°E, S62°E, N20°W.
- Azimuth: The clockwise angle from the north branch of a meridian, e.g. 45° (northeast), 180° levelling is more accurate since elevation differences (south).
- Deflection angle: The angle between a line and the prolongation of a preceding line; it is a right or left angle depending on whether the new line is right (clockwise) or left (counterclockwise) of the preceding line.
- Interior angle: An angle inside a closed polygon.

Types of Traverses

- Azimuth: Along a single direction (azimuth); common for slope profile surveys where the profile is always perpendicular to contours (i.e. maximum slope angle).
- Closed traverse: Begins and ends at fixed control points of known location; permits calculation and adjustment for closure error.
- Closed-loop traverse: Begins and ends at the same station; permits calculation and adjustment for closure error and use of interior and deflection angles.
- Open traverse: Surveying from a known position to a point of unknown position; does not enable computational checks for error, rather all measurements must be repeated to check for error.

Shape of the Earth and Error

The two basic problems in topographic surveying:

The shape of the earth (the geoid) is a consideration only in geodetic surveying, where over long distances flat surfaces are not level, plumb lines are not parallel and the sum of the angles in a triangle is greater than 180°; thus precise surveys over large areas employ the principles of geodesy (the mathematical properties of an ellipsoid that emulates the earth); however with most surveys, including virtually all topographic surveying in geography, the departure of horizontal lines from level and plane angles from spherical angles are negligible and can be ignored.

Plane Surveying

Where the earth's surface is regarded as a plane; level lines are considered straight, angles are considered to be plane angles and plumb lines are considered to be parallel within the survey with these assumptions, the relative locations of points can be calculated using the principles of plane trigonometry.

For right angles: $\text{Sin A} = a/c$, $\sin B = b/c$, and $c^2 = \left(a^2 + b^2\right)$.

Where,

A & B are the acute angles.

C is the right angle and a, b & c are the sides opposite the angles designated with the same letter. Thus the sides and angles can be calculated with a knowledge of 1 side and 1 angle or 2 sides.

For oblique triangles: $a/\sin A = b/\sin B = c/\sin C$; thus given 2 angles and one side or 2 sides and 1 angle are the other sides and angles can be calculated.

Error

Distances and angles can never be determined exactly; measurements are subjected to error. Error can be controlled through procedure and instrumentation. Surveys are conducted according to standard levels of accuracy (first order, second order, etc). The desired level of accuracy depends on the intended survey data (e.g. locating permanent stations or surveying bridges and dams versus surveying for terrain analysis or orienteering).

For topographic mapping, the desired level accuracy is the plottable error, the shortest distance that can be depicted on a map at a given scale. The drafting of lines generally is accurate to within 0.25 mm. At 1:1000, 0.25 = 250 mm or 0.25 m on the ground. Optical measuring devices will provide this level of accuracy. At 1:25,000, 0.25 mm = 6.25 m on the ground. Pacing of distance will provide this level of accuracy, although in practice accuracy is greater than the plottable error by as much as one-third (e.g. 80 mm rather than 250 mm at a scale of 1:250,000) so that plotting and surveying errors are not compounded.

Adjusting for Closure Error

Horizontal Angles

In a closed polygon, the sum of the interior angles = 180° (n-2), where n is the number of sides in the polygon, thus the sum of the horizontal angles in a triangle (n = 3) is 180°; an equal angle is subtracted or added to each measurement to satisfy the equation for interior angles; if the closure error is not equally divisible by n, make the largest adjustments to the largest angles.

The sum of deflection angles for any closed polygon is always 360°; this provides for another means of determining and adjusting for closure error.

Difference in Elevation

Closure error can be determined for closed and closed-loop traverses. The closure error can be divided by the number of stations on the traverse or the correction at each station can be calculated according to the distance from the origin of the survey.

$$Ci = di / L * E_c,$$

Where,

Ci = The correction applied to station I.

di = The distance to station i from the origin of the traverse.

L = The total length of the traverse.

E_c = The closure error.

This method accounts for the propagation of error with distance.

Horizontal Distances

As with levelling, closure error can be determined for closed and closed-loop traverses, where the coordinates of the end points are identical or known. Location in the horizontal plane are given by x and y coordinates (e.g. northing and easting). Using the measured horizontal distances and adjusted angles, calculate the co-ordinates of each station. The difference between the calculated and known co-ordinates of the end control point is dx and dy, the closure error in x and y. As with levelling, the adjustment is a function of the distance traversed (Li) relative to the total length of the traverse (L).

$$Cdx_i = dx * L_i / L \; Cdy_i = dy * L_i / L$$

Where, Cdx_i and Cdy_i are the adjustments in x and y coordinates at station i.

The relative accuracy of distance measurements can be expressed as $((dx^2 + dy^2)1/2)/L$. An angular error of one minute is equivalent to a distance measurement error of 3 cm over a distance of 100 m, since the sine of 1/60° is .00029.

The vertical direction is parallel to the direction of gravity; at any point, it is the direction of a freely suspended plumb-bob cord. The vertical distance of a point above or below a given reference surface is called the elevation of the point. The most commonly used reference surface for vertical distance is mean sea level. The vertical distances are measured by the surveyor in order to determine the elevation of points, in a process called running levels or levelling. The determination and control of elevations constitute a fundamental operation in surveying and engineering projects. Levelling provides data for determining the shape of the ground and drawing topographic maps and the elevation of new facilities such as roads, structural foundations, and pipelines.

Finding Vertical and Horizontal Elevation Differences between Points

There are two common methods of surveying to determine slope distance. The traditional method involves stadia, where the difference in rod readings at the upper and lower stadia hairs are converted to a slope distance. The newer method uses a total station instrument and prism pole to determine the slope distance electronically.

The two methods of acquiring the slope distance require different methods of determining the vertical and horizontal distances between the instrument and the point of interest. The total station and prism pole gives a true slope distance. However, the stadia method requires an extra correction factor due to an inclined line of sight reading on a vertical rod. When hand calculating the vertical distance between points, this difference needs to be taken into consideration. Because the total station gives a true slope distance, the vertical difference is simply written as:

$$V = s * \sin \alpha$$

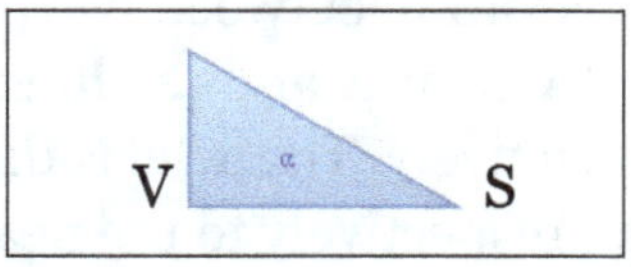

However, with the stadia reading a correction for the angle between the line of sight and the rod needs to be included.

$$V = s * (\sin 2\alpha)/2$$

When calculating the horizontal distance between points, this difference also needs to be taken into consideration. Because the total station gives a true slope distance, the horizontal difference is simply written as,

$$H = s * \cos \alpha$$

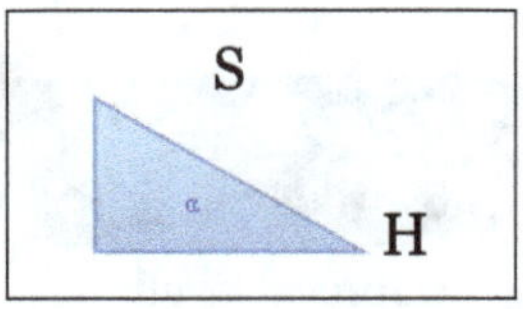

However, with the stadia reading a correction for the angle between the line of sight and the rod needs to be included.

$$H = s * \cos \alpha * \cos \alpha = s * \cos 2\alpha$$

This difference between total station measurements and stadia readings is also accounted for in survey note reduction programs such as SSRP and Auto CAD. These programs usually ask whether slope distances are from stadia or prisms. The program then applies the correct formulas. It is important that this variable is correct in the reduction program we are using.

2.5 Reciprocal Levelling

A method of differential levelling is based on the fact that the instrument is kept equidistant from the back and forward station. The difference in elevation of two stations is equal to the difference of the staff readings. By setting the level midway the error, due to the curvature and refraction and also the collimation error is eliminated.

Applications

- To find accurate relative elevations of two widely separated inter-visible points (between which levels cannot be set), reciprocal levelling is being used.
- To find the difference in elevation between two points, say X and Y (as shown in figure), a level is set up at L near X and readings (X_1 and Y_1) are observed with staff on both X and Y respectively. The level is then set up near Y and staff readings (Y_2 and X_2) are taken respectively to the near and distant points.
- If the differences in the set of observations are not same, then the observations are fraught with errors. The errors may arise out of the curvature of the earth or intervening atmosphere (associated with variation in temperature and refraction) or instrument (due to error in collimation) or any combination of these.

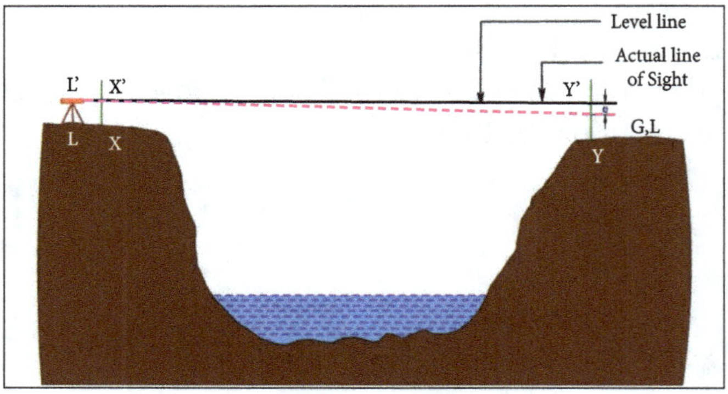

Reciprocal levelling.

The true difference in elevation and errors associated with observation, if any, can be found as follows.

Let the true difference in elevation between the points be r_h and the total error be e. Assuming, no error on observation of staff near the level (as the distance is very small).

Then,

$$r_h = X_1 \sim (Y_1 - e) \text{ [From first set of observation].}$$

$$r_h = (X_2 - e) \sim Y_2 \text{ [From second set of observation].}$$

or,.

$$\Delta h = \frac{(X_1 \sim Y_1) + (X_2 \sim Y_2)}{2}$$

or,

$$\Delta h = \frac{\text{First difference in observation} + \text{Second difference in observation}}{2}$$

Thus, the true difference in elevation between any two points can be obtained by taking the mean of the two differences in observation.

$$\text{Total error}, e = \frac{(X_1 \sim Y_1) \sim (X_2 \sim Y_2)}{2}$$

or,

$$e = \frac{\text{First difference in observation} - \text{Second difference in observation}}{2}$$

Thus total error in observations can be obtained by taking the difference of the two differences in observation. The total error consist of error due to curvature of the earth, atmospheric errors (due to temperature and refraction) and instrumental errors (due to error in collimation) etc.

3

Mapping and Sensing

3.1 Mapping Details and Contouring

A line joining points of equal elevations is called a contour line. It facilitates depiction of the relief of terrain in a two dimensional plan or map.

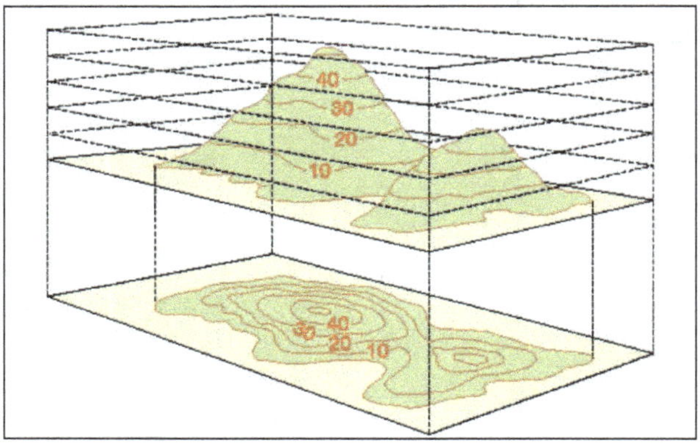

Levelling and contouring.

Contour Interval

The difference in elevation between successive contour lines on a given map is fixed. This vertical distance between any two contour lines in a map is called the contour interval (C.I) of the map. The first figure shows contour interval of 1m whereas, the second figure shows 10m.

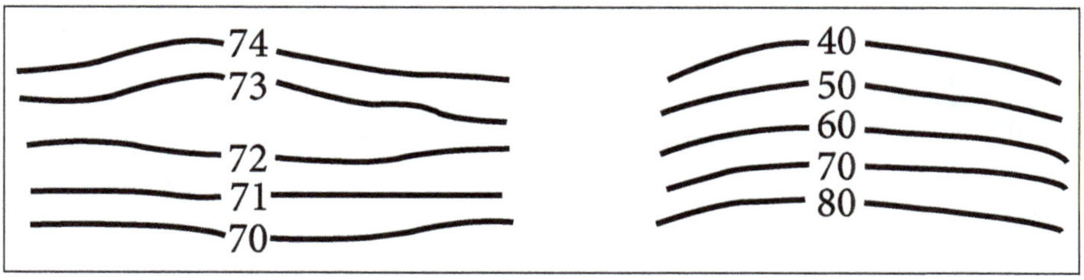

Contour lines.

The choice of suitable contour interval in a map depends upon four principal considerations. These are:

- Nature of the terrain.

- Scale of the map.
- Accuracy.
- Time of cost.

Contour

Contouring is the science of representing the vertical dimension of the terrain on a two dimensional map. We can understand contouring by considering a simple example.

Let us assume that a right circular cone of base 5m diameter and vertical height 5m is standing upright on its base. Let the base be resting on a horizontal plane at zero level as shown in below figure:

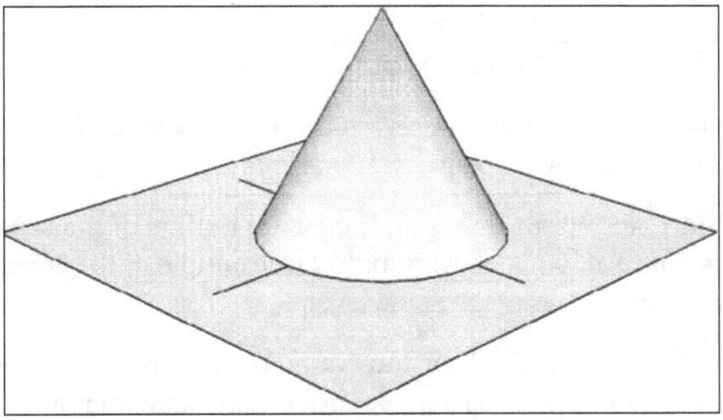

Cone on a 0m level plane.

At zero level, the outline of the cone will be a circle of 5m diameter. This circle is the contour line at 0m elevation for the cone. We draw this first contour line on paper to a convenient scale.

Let us now slice the cone at 1m height from the base. This will produce another circular outline corresponding to the diameter of the cone at 1m elevation. Let us draw this second circle on our contour map using the same scale. The second circle being smaller in diameter than the first will appear as a concentric circle within the first circle.

Similarly, we continue to draw the outline of the cone at 2m, 3m, 4m and 5m levels on our contour map. Our contour map for the conical object is now ready. The circles on the map are called contour lines.

Like the cone in our example, hills project upwards from ground level. The contour map of a hilly terrain will be similar to that of the cone, except that instead of perfect circles, the contour lines would be of irregular shapes. The important point of similarity to note here is that hilly terrain would be represented by contour lines with increasing elevation towards the centre.

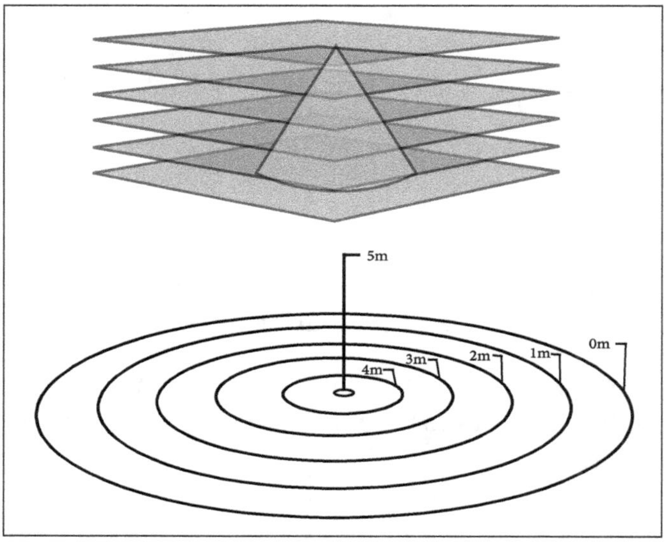

Cone sliced at 1m intervals (top). Contour map of the cone drawn to a convenient scale (bottom).

In contrast to this, a pond or depression would be represented by contour lines with decreasing elevation towards the centre.

- Contour line: A Contour line is an imaginary outline of the terrain obtained by joining its points of equal elevation. In our example of the cone, each circle is a contour line joining points of same level.
- Contour Interval (CI): Contour interval is the difference between the levels of consecutive contour lines on a map. The contour interval is a constant in a given map. In our example, the contour interval is 1m.
- Horizontal Equivalent (HE): Horizontal equivalent is the horizontal distance between two consecutive contour lines measured to the scale of the map.

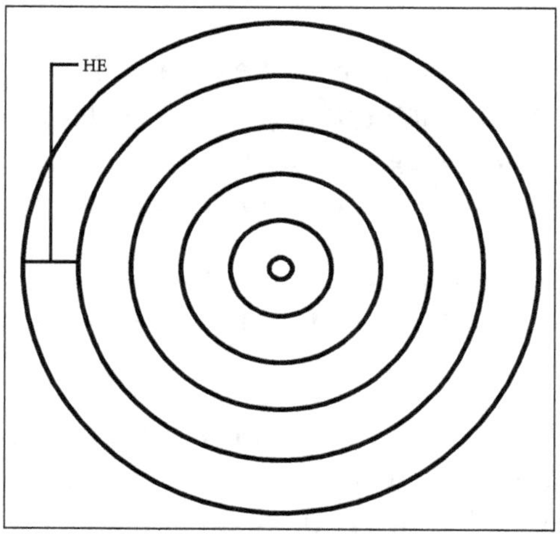

Horizontal equivalent.

- Gradient: Gradient represents the ascending or descending slope of the terrain between two consecutive contour lines. The slope or gradient is usually stated in the format 1 in S, where 1 represents the vertical component of the slope and S its corresponding horizontal component measured in the same unit.

The gradient between two consecutive contour lines can also be expressed in terms of Tan Q θ as follows. Tan Q θ = CI / HE ... both measured in the same unit.

Difference between Contour Interval and Horizontal Equivalent

There are three main differences between contour interval and horizontal equivalent as follows:

S. No.	Contour interval	Horizontal equivalent
1.	It is based on vertical levels.	Represents horizontal distance.
2.	No measurement or scaling is required since the contour levels are indicated on the contour lines.	The distance must be measured on the map and converted to actual distance by multiplying with the scale of the map.
3.	In a given map the contour interval is a constant.	The horizontal equivalent varies with slope. Closer distance indicates steep slope and wider distance represents gentle slope.

Factors Governing Selection of Contour Intervals

The survey leader has to decide an appropriate contour interval for his project before start of survey work. The following factors govern the selection of contour interval for a project.

S.No	Factor	Select High CI like 1m, 2m, 5m or more	Select Low CI like 0.5m, 0.25m, 0.1m or less
1.	Nature of ground	If the ground has large variation in levels, for instance, hills and ponds.	If the terrain is fairly level.
2.	Scale of the map	For small scale maps covering a wide area of varying terrain.	For large scale maps showing details of a small area.
3.	Extent of survey	For rough topographical map meant for initial assessment only.	For preparation of detailed map for execution of work.
4.	Time and resources available	If less time and resources are available.	If more time and resources are available.

Characteristics of Contour

Contours show distinct characteristic features of the terrain as follows:

- All points on a contour line are of the same elevation.

- No two contour lines can meet or cross each other except in the rare case of an overhanging vertical cliff or wall.
- Closely spaced contour lines indicate steep slope.
- Widely spaced contour lines indicate gentle slope.
- Equally spaced contour lines indicate uniform slope.
- Closed contour lines with higher elevation towards the centre indicate hills.
- Closed contour lines with reducing levels towards the centre indicate pond or other depression.
- Contour lines of ridge show higher elevation within the loop of the contours. Contour lines cross ridge at right angles.
- Contour lines of valley show reducing elevation within the loop of the contours. Contour lines cross valley at right angles.
- All contour lines must close either within the map boundary or outside.

Plotting Methods of Interpolating Contours

A contour plot is a graphical technique for representing a 3-dimensional surface by plotting constant z slices, called contours, on a 2-dimensional format. That is, given a value for z, lines are drawn for connecting the (x, y) coordinates where that z value occurs.

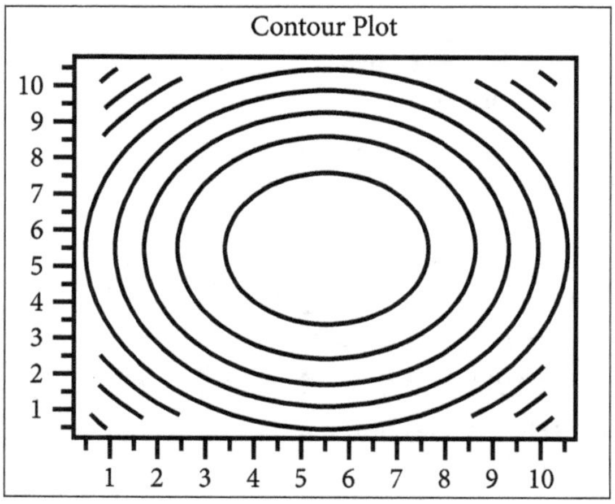

Contour plot.

This contour plot shows that the surface is symmetric and peaks in the center.

The contour plot is formed by:

Vertical axis: Independent variable 2.

Horizontal axis: Independent variable 1.

Lines: Iso-response values.

The independent variables are usually restricted to a regular grid. The actual techniques for determining the correct iso-response values are rather complex and are almost always computer generated.

An additional variable may be required to specify the Z values for drawing the iso-lines. Some software packages require explicit values. Other software packages will determine them automatically.

If the data (or function) do not form a regular grid, we typically need to perform a 2-D interpolation to form a regular grid. The contour plot is an alternative to a 3-D surface plot.

Methods of Interpolating Contours: Direct Methods

1. Vertical Control

In this method, a benchmark is required in the project area. The level is set up on any commanding position and back sight is taken on the bench mark. Let the back sight reading on the bench mark be 1.485 m.

If the reduced level of the bench mark is 100 m, the height of instrument would be 100 + 1.485 = 101.485 m.

To locate the contour of 100.5 m value, the staff man is directed to occupy the position on the ground where the staff reading is 101.485 -100.500=0.985m.

Mark all such positions on the ground where the staff reading would be 0.985 m by inserting pegs. Similarly locate the points where the staff reading would be 101.485 -101 = 0.485 m for 101m contour.

The contour of 101.5 m cannot be set from this setting of the instrument because the height of instrument for this setting of the instrument is only 101.485 m.

Therefore, locating contours of higher value, the instrument has to be shifted to some other suitable position. Establish a forward station on a firm ground and take fore sight on it. This point acts as a point of known elevation, for shifting the position of the instrument to another position, from where the work proceeds in the similar manner till the entire area is contoured.

2. Horizontal Control

Horizontal control surveys co-ordinate horizontal positional data. These positions can be referenced by parallels or plane co-ordinate axes. Because they are used as a framework for other surveys, these surveys must be precise and accurate.

These surveys provide a network of monuments or points on the ground that can be used as the control for any other surveying project, such as a boundary or construction survey.

The advantages of using a horizontal control survey is that lost monuments can be replaced accurately, surveys can be co-ordinated, more than one network station can provide a check to the work, and a reduction in the cost of the project can be achieved. Most horizontals should (and will) be connected to the control network.

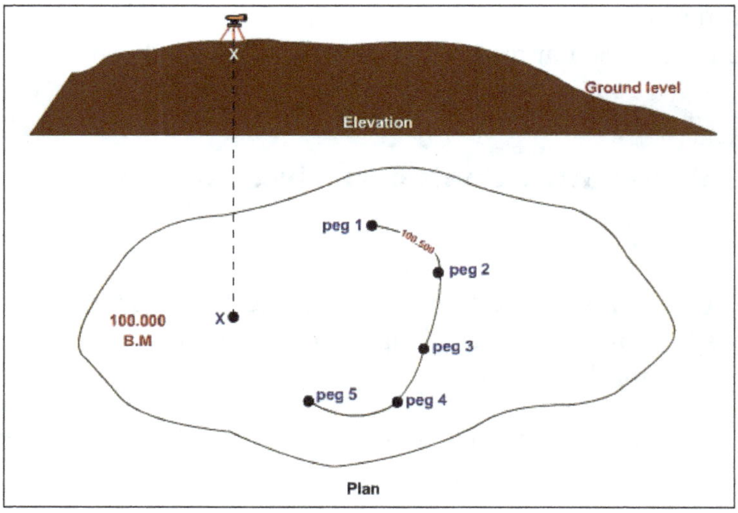

Horizontal Control.

3. Radial Method

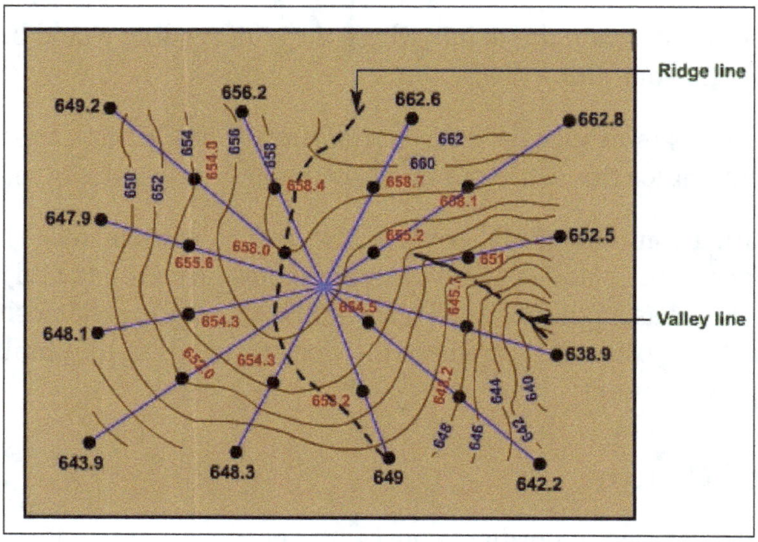

Radial Method.

In this method, a number of radial lines are set out at known angular interval at each station and points are marked at the ground at convenient distance apart on the rays that are set. Spot levels of these points are determined by levelling.

The points are plotted to the scale of the map and spot levels are entered. The contours of desired values are then located by interpolation. This method is convenient in hilly terrain with level stations chosen at high points so as to command a large area from each. Horizontal control may be obtained by taping.

Methods of Interpolating Contours: Indirect Methods

1. Cross Section Method

Cross section method is most suitable for preparing contour maps for road works, rail works, canals etc. Typically, this type of land has a very long strip but narrow width.

The steps involved are as follows:

- The centre line of the strip of land is first marked.
- Lines perpendicular to the longitudinal strip are marked dividing the strip into equal sections.
- The perpendicular lines are divided into equally spaced divisions, thus forming rectangular grids.
- Levels are taken at the intersection of the grid lines to obtain the cross-section profile of the strip of land.
- Contour map is plotted in the office by interpolating points of equal elevation based on the levels taken at site.

2. Squares or Grid Method

Squares or grid method is suitable for contouring of plains or gently sloping grounds.

The steps adopted are as follows:

- Mark square grids on the land to be surveyed. The grid size would depend on the extent of survey. Generally a 1m x 1m grid is selected for small works and a larger grid size for large works.
- Levels are taken at all the corners of the square and the intersection of the diagonal.
- Levels taken on the intersection of diagonals is used for verification of the interpolation.

- Contour map is plotted in the office by interpolating points of equal elevation based on the levels taken on the corners of the square.

3. Tacheometric Method

Tacheometric method is adopted for contouring of very steep hills.

The steps are as follows:

- Set up the tacheometer at the top of the steep hill. Tacheometer is a theodolite fitted with stadia diaphragm. The stadia diaphragm has three horizontal parallel hairs instead of one as found in a conventional cross hair diaphragm.
- With the help of a tacheometer it is possible to determine the horizontal distance of the point from the telescope as well its vertical level.
- The steep hill is surveyed at three levels – the base of the hill, the mid-level of the hill and the top level of the hill.
- Using the tacheometer, readings are taken all around the hill at equal angular intervals on all these three levels.
- The radial plot thus obtained is worked in the office to interpolate points of equal elevation for contour mapping.

3.2 Profile Cross Sectioning and Measurement of Areas and Volumes

Profile Cross Sectioning

Cross section method is most suitable for preparing contour maps for road works, rail works, canals etc. Typically, this type of land has a very long strip but narrow width.

The steps involved are as follows:

- The centre line of the strip of land is first marked.
- Lines perpendicular to the longitudinal strip are marked dividing the strip into equal sections.
- The perpendicular lines are divided into equally spaced divisions, thus forming rectangular grids.
- Levels are taken at the intersection of the grid lines to obtain the cross-section profile of the strip of land.

- Contour map is plotted in the office by interpolating points of equal elevation based on the levels taken at site.

Profile Levelling

The process of determining the elevations of a series of points at measured intervals along a line such as the centerline of a proposed ditch or road or the centerline of a natural feature such as a stream bed. Normally we will assign an elevation of 100.00 to the datum rather using the mean sea level elevation.

An Extension of Differential Levelling

- Elevations are determined in the same manner.
- The same definitions define the concepts and terms involved.
- The same types of mistakes and errors are possible.
- A page check should always be done.
- A closure check should be done if the profile line runs between bench marks.

Accuracy of Rod Readings

The back sights, foresights, and elevations of benchmarks and turning points should be recorded to the nearest 0.01 ft. Profile elevations of intermediate points are determined from "ground readings" and thus the foresight readings and subsequent elevations should be recorded to the nearest 0.1 ft.

Theory

- Add rod readings (BS) to benchmark or known turning point elevations to get the elevation of the line of sight (HI).
- Subtract rod readings (FS) from the line of sight to establish elevations of unknown points.
- Take any number of intermediate FS readings at points along the line until it is necessary to establish a turning point to move the level.
- Repeat as required.

Location of Intermediate Points

- A foresight is taken on a bench mark to establish the height of instrument.
- A foresight is taken on the stations as required (such as every 100 ft.).

- Foresights are also taken at breaks in the ground surface and at critical points.
- This is repeated until the limit of accurate sighting is reached, at a point at which a turning point is established and the level is moved.

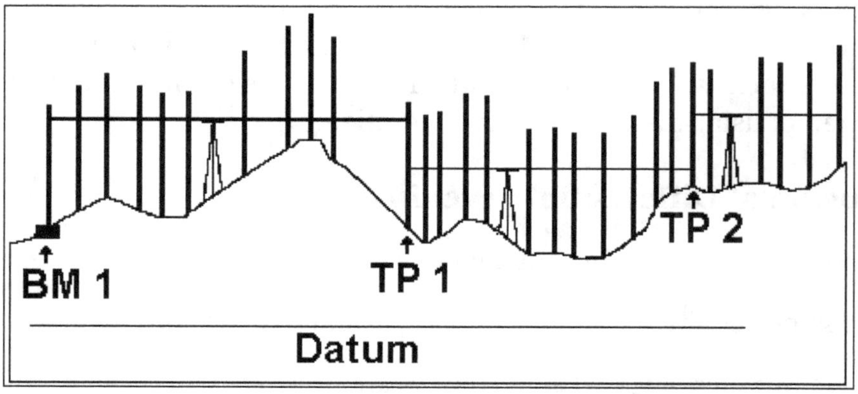

Schematic.

The level is usually set up off the center line.

Profile Cross Sections

Cross sections are lines of levels or short profiles made perpendicular to the center line of the project. (For example, taking a cross section profile of a stream bed while doing a profile survey of the stream.)Cross sections are usually taken at regular intervals and at sudden changes in the center-line profile.

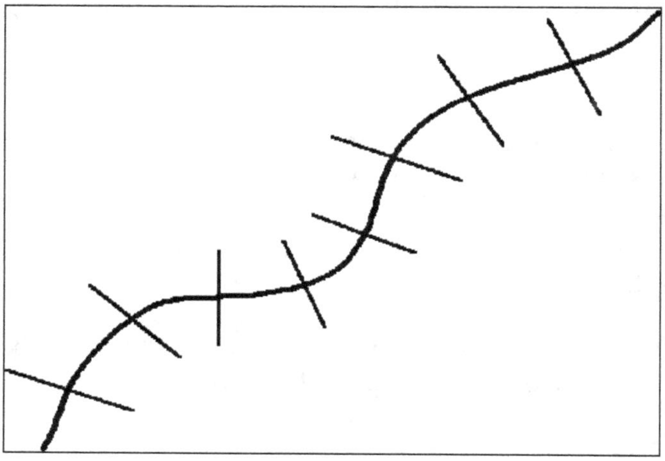

Cross section schematic.

Cross Sections

- The cross sections must extend a sufficient distance on each side of the center line to provide a view of the surrounding terrain.

- Rod readings should be taken at equal intervals on both sides of the center line and at significant changes in the terrain.

Example: For a stream cross section, rod readings could be taken at 15, 30, 45, and 60 ft. on each side of the center line as well as the edge of the stream and the top of bank of the stream.

Cross Section Field Notes: Field notes for a cross section should include, an elevation or difference in elevation from the center line and horizontal distance from the center line.

Measurement of Areas and Volumes

Calculation of area is carried out by any one of the following methods:

1. Mid-ordinate Method

The rule states that if the sum of all the ordinates taken at midpoints of each division multiplied by the length of the base line having the ordinates (9 divided by number of equal parts).

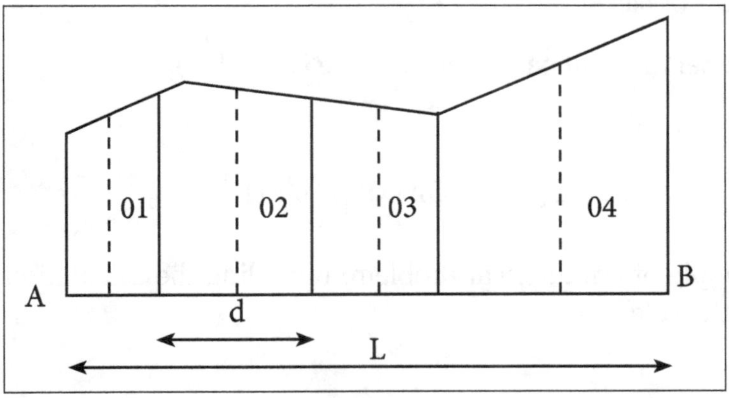

Mid ordinate rule.

In this, base line AB is divided into equal parts and the ordinates are measured in the midpoints of each division.

$$\text{Area} = \left(\left[O_1 + O_2 + O_3 + \ldots.. + O_n\right] * L\right) / n$$

Where,

L = length of baseline.

n = number of equal parts, the baseline is divided.

d = Common distance between the ordinates.

2. Average Ordinate Method

The rule states that (to the average of all the ordinates taken at each of the division of equal length multiplies by baseline length divided by number of ordinates).

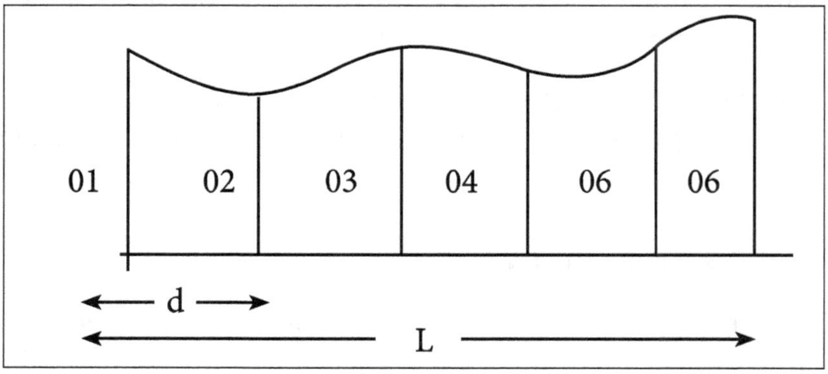

Average ordinate rule.

O_1, O_2, O_3, O_4....On ordinate taken at each of division.

Where,

L = length of baseline.

n = number of equal parts (the baseline is divided).

d = Common distance.

$$\text{Area} = \left[(O1 + O2 + O3 + \ldots + On) * L\right] / (n+1)$$

Here is an example of a numerical problem, regarding the calculation of areas using Average Ordinate Rule.

Problems

1. The following perpendicular offsets were taken at 10m interval from a survey line to an irregular boundary line.

9, 12, 17, 15, 19, 21, 24, 22, 18

Let us calculate area enclosed between the survey line and irregular boundary line.

Solution:

Given data:

Offset value = 9, 12, 17, 15, 19, 21, 24, 22, 18

Interval = 10m

Formula:

$$\text{Area} = \left[\left(O_1 + O_2 + O_3 + \ldots. + O_9\right) * L\right] / (n+1)$$

$$= \left[(9+12+17+15+19+21+24+22+18) * 8 * 10\right] / (8+1)$$

$$\text{Area} = 139538\text{sqm}.$$

3. Trapezoidal Rule

In this method, boundaries between the ends of ordinates are assumed to be straight. Thus, the area enclosed between these line and the irregular boundary lines are considered as trapezoids.

$$A = d/2(O1 + On + 2(O2 + O3 + O4 + \ldots\ldots\ldots. + On - 1))$$

A = distance between ordinate/ 2*[{first ordinate + last ordinate} + 2{sum of other ordinates).

Problems

1. The following offsets were taken from a chain line to an irregular boundary line at an interval of 10 m.

0, 2.50, 3.50, 5.00, 4.60, 3.20, 0 m.

Let us calculate the area between the chain line, the irregular boundary line and the end offsets by Trapezoidal Rule.

Solution:

Given data:

Offset value = 0, 2.50, 3.50, 5.00, 4.60, 3.20, 0 m

Interval = 10m

Formula:

$$A = d/2\left(O_1 + O_n + 2\left(O_2 + O_3 + O_4 + \ldots\ldots\ldots. + O_{n-1}\right)\right)$$

Here,

$$d = 10/2\ (0 + 0 + 2(2.50 + 3.50 + 5.00 + 4.60 + 3.20)$$

$$\text{Area} = 5 * 37.60 = 188\ \text{m}^2.$$

4. Simpson's Rule

It states that, sum of first and last ordinates has to be done. Add twice the sum of remaining odd ordinates and four times the sum of remaining even ordinates. Multiply to this total sum by 1/3rd of the common distance between the ordinates which gives the required area.

Where,

O_1, O_2, O_3,.... On are the lengths of the ordinates.

d = Common distance.

n = Number of divisions.

Note: This rule is applicable only if ordinates are odd, i.e. even number of divisions.

Volume: Volumes are computed from cross-section measurements by the average end area method.

Volume (yd^3) = [L x (A_1 + A_2)] (2 x 27) L is in feet A_1 and A_2 are in square feet Volume (metric) (m^3) = [L x (A_1 + A_2)] 2 L is in meters A_1 and A_2 are in square meters.

3.3 Application of Measurements in Quantity Computations

Quantity computations are commonly performed by computer, but hand estimating quantities is always prudent for the purpose of verifying computer performed quantities, and it is not uncommon for the QA inspector to be tasked with manually performing other quantity computations. For example, excavation within the specified payment limits may have been computed before it was learned that additional excavation would be required. In this case, the inspector may be tasked with computing the quantity of any additional excavation beyond the specified payment limits. In doing this, the inspector may survey or oversee the survey required to obtain the data needed to compute the additional quantities, verify the adequacy and accuracy of the survey and resulting data, and perform the computations necessary to determine the quantity of additional excavation. For this and similar tasks, a fundamental knowledge of quantity computations is needed. Even if someone else performs all of the computations, a fundamental knowledge of how to perform quantity computations is needed for the inspector to verify the survey data is adequate for making quantity computations.

There are two categories of quantity computations made in construction work:

- Interim quantity computations.
- Final quantity computations.

Interim quantity computations are made to estimate quantities for progress payments.

Final quantity computations are made to compute final quantities for payment. Interim computations do not have to be as detailed or precise as final quantity computations, but the inspector should verify they are accurate enough to provide a reasonable estimate of the quantity of completed work.

To ensure, adequate survey data will be obtained for final quantity computations, the computations should be set up or planned in detail, prior to obtaining the survey data. The person who will perform the final quantity computations should initiate the computation process by selecting the computation method and laying out computations to the point where the amount of detail and required format of the field survey data is determined. This should be done before or near the beginning of construction since quantity surveys are begun early in the job at or near the time of staking.

The necessary details and presentation formats should be conveyed and should be understood by the construction inspector and surveyor prior to beginning the surveying for final quantity computations. The limits of payment should be well defined in the specifications or otherwise clearly documented prior to the beginning of the work. The inspector should become familiar with the specified limits and verify that quantity surveys and computations are consistent with these limits. For example, the payment limits for earth fill may be defined as the lower limits of excavation and the neat line and grade of the earth fill. In this case, the inspector would verify that survey data adequately, defining the lower limits of excavation is obtained for computing the earth fill quantity.

Volume, area, and linear computations are made to compute quantities for payment. Some items, such as rock riprap, are sometimes paid on a weight basis, which requires a summary of the delivery ticket weights to compute the quantity for payment. All computations should be performed accurately and presented in a logical order. They must be checked by someone other than the person performing the computations. Computations are not considered to be complete until they have been checked.

The inspector's responsibilities related to quantity computations include:

- Maintaining a fundamental knowledge of quantity computations.
- Hand estimating to verify computer generated quantity values.
- Verifying interim computations are accurate enough to provide a reasonable estimate of the quantity of completed work.
- Verifying the necessary detail and presentation formats of quantity surveys is understood by the surveyor prior to beginning quantity computation surveys.
- Verifying contractor performed quantity computations are consistent with specified payment limits.

- Verifying all items to be submitted by the contractor are submitted within the specified time frame.

3.3.1 Survey Station

Survey stations are of two kinds:

- Main stations: Main stations are the end of the lines, which command the boundaries of the survey and the lines joining the main stations are called the main survey line or the chain lines.
- Subsidiary or the tie stations: Subsidiary or the tie stations are the point selected on the main survey lines, where it is necessary to locate the interior details such as fences, hedges, building etc.

3.4 Introduction of Remote Sensing and its Applications

It is the science of acquiring information about the Earth's surface without actually being in contact with it. This is done by sensing and recording reflected or emitted energy and processing, analyzing and applying that information.

In much of remote sensing, the method involves an interaction between incident radiation and the targets of interest. This is exemplified by the use of imaging systems where the following seven elements are involved. Note, but that remote sensing also involves the sensing of emitted energy and the use of non-imaging sensors.

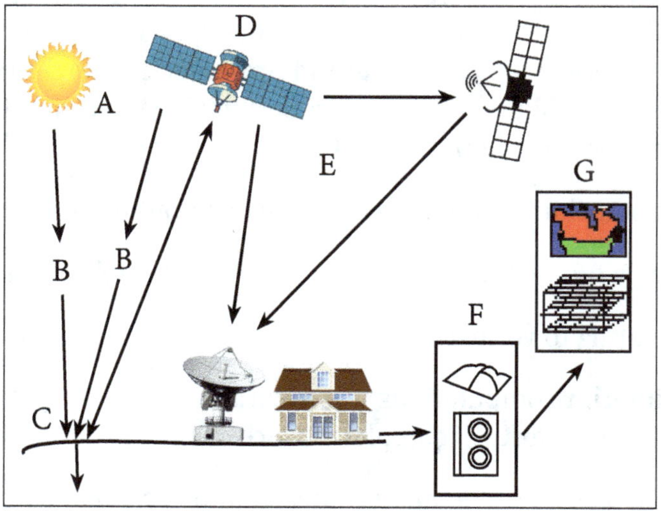

- Energy Source or Illumination (A): The first requirement for remote sensing is to have an energy source that illuminates or gives electromagnetic energy to the target of interest.

- Radiation and the Atmosphere (B): As the energy travels from its source to the target, it can come in contact with and interact with the atmosphere it passes through. This interaction may take place a second time as the energy travels from the target to the sensor.
- Interaction with the Target (C): Once the energy makes its way to the target through the atmosphere, it interacts with the target depending on the properties of both the target and the radiation.
- Recording of Energy by the Sensor (D): After the energy has been scattered by or emitted from the target, we need a sensor to collect and record the electromagnetic radiation.
- Transmission, Reception, and Processing (E): The energy recorded by the sensor has to be transmitted, often in electronic form, to a receiving and processing station where the data are processed into an image.
- Interpretation and Analysis (F): The processed image is interpreted, visually or digitally or electronically, to extract information about the target which was illuminated.
- Application (G): The final element of the remote sensing process is achieved when we apply the information we have been able to extract from the imagery about the target in order to better understand it, reveal some new information or assist in solving a particular problem.

Applications of Remote Sensing

The era of satellite remote sensing was open in the 1960s when cameras and electronic sensors were mounted on spacecraft. These days there is a big assortment of satellite systems actively recording information regarding the Earth. A wide variety of imagery is available from satellites. Both active and passive sensors, operating from the microwave to the ultraviolet regions of the electromagnetic spectrum collect a large amount of information regarding the earth's surface every day. Each of the systems vary in terms of their spatial, spectral, radiometric and temporal resolution. These characteristics play an important role in defining which applications the sensor is best suited for. The main benefits of satellite remote sensing are the following:

- The data are available for large areas: for example: 35000km^2 for LANDSAT scene, 3600 km^2 for SPOT scene.
- They are available on a regular basis for all points on the globe: data may be acquired every 1-3 days.
- They are objective: the sensor-transmission-reception system involves no human intervention.

The Data Collected are Related to the Earth Surface Features

To collect data the sensors use the wide spectrum of electromagnetic spectrum and use several band at once. They are in digital form and geometrically corrected images will be used to provide a base to overlay other data or to be used as part of an analysis in a GIS environment.

The applications of remote sensing summarized in this unit are only representative, but not exhaustive. We focus on applications tied to the Earth surface. To show the remote sensing as data source and the benefits of remote sensing applications we stressed some examples of touched fields.

Agricultural

Agriculture plays a vital role in economies of countries. The production of food is important to everyone and producing food in a cost-effective manner is the goal of every farmer and an agricultural agency. The satellites has a capability to image individual fields, regions and counties on a frequent revisit cycle. Customers will receive field-based information including crop identification, crop area determination and crop condition monitoring. Satellite data are employed in precision agriculture to manage and monitor farming practices at various levels. The data can be used to farm optimization and spatially-enable management of technical operations. The images will help determine the location and extent of crop stress and then will be used to develop and implement a spot treatment plan that optimizes the use of agricultural chemicals. The major agricultural applications of remote sensing include the following:

- Vegetation.
- Crop type classification.
- Crop condition assessment.
- Crop yield estimation.
- Soil.
- Mapping of soil characteristics.
- Mapping of soil type.
- Soil erosion.
- Soil moisture.
- Mapping of soil management practices.
- Compliance monitoring.

Crop Type Classification

Remote sensing technology will be used to prepare maps of crop type and delineating their extent. Traditional methods of obtaining this information are census and ground surveying. The use of satellites is advantageous as it can generate a systematic and repetitive coverage of a large area and provide information regarding the health of the vegetation. The data of crop is needed for agricultural agencies to prepare an inventory of what was grown in certain areas and when. This information serves to predict grain crop yield, collecting crop production statistics, mapping soil productivity, facilitating crop rotation records, identification of factors influencing crop stress, assessment of crop damage and monitoring farming activity.

Different Kind of Images can be Used

There are different types of remote sensing systems used in agriculture but the most common is a passive system that senses the electromagnetic energy reflected from plants. The spectral reflection of a vegetation depend on stage type, changes in the phenology and crop health, and thus can be measured and monitored by multi-spectral sensors. Several remote sensing sensors operate in the green, red and near infrared regions of the EM spectrum, they measure both absorption and reflectance effects associated with vegetation. Multi-spectral variations facilitate fairly precise detection, identification and monitoring of vegetation. The observation of vegetation phenology requires multi-temporal images. Various sensors often provide complementary information, and when integrated together, will facilitate interpretation and classification of imagery. Examples include combining high resolution panchromatic imagery with coarse resolution multi-spectral imagery or merging actively and passively sensed data.

Some species (red color) of cultivated plant on satellite image.

Crop Monitoring and Damage Assessment

Remote sensing has a number of attributes that lend themselves to monitoring the health of crops. The optical sensing advantage is that it will see the infrared, where wavelengths are highly sensitive to crop vigor as well as crop stress and crop damage. Remote sensing imagery also gives the needed spatial overview of the land. Remote sensing can aid in identifying crops affected by conditions that are too dry or wet, affected by insect, weed or fungal infestations or weather related damage. Images will be obtained throughout the growing season to not only detect problems, but also to monitor the success of the treatment. Detecting damage and monitoring crop health requires high-resolution, multi-spectral imagery and multi-temporal imaging capabilities. One of the most critical factors in making imagery useful to farmers is a quick turnaround time from data acquisition to distribution of crop information.

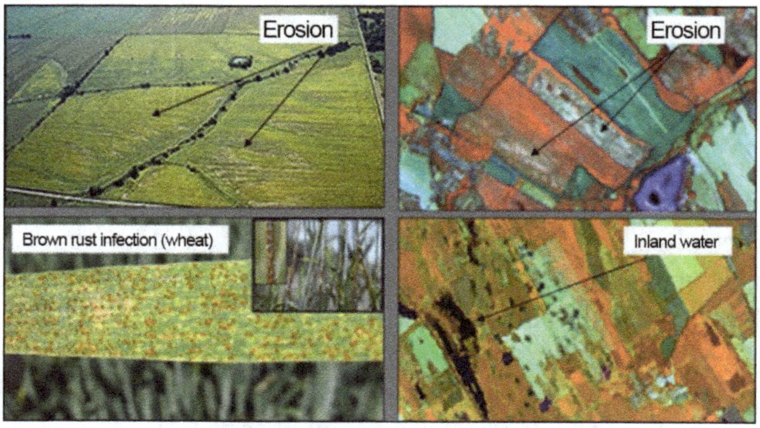

The problems inside the agricultural fields.

Soil Mapping

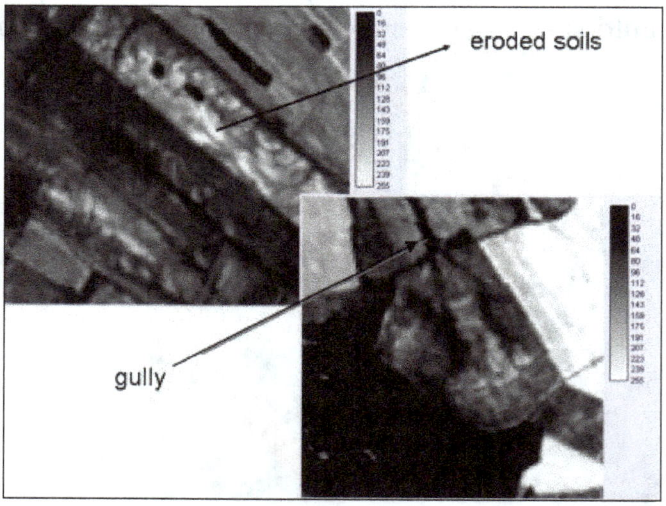

Some kind of erosion in satellite image.

The disturbance of soil by land use impacts on the quality of our environment. Salinity, soil acidification and erosion are some of the issues. Remote sensing is a good method for mapping and prediction of soil degradation. Soil layers that rise to the surface during erosion have different color, tone and structure than non-eroded soils thus the eroded parts of soil will be easily identify on the images. Using multi-temporal images we will study and map dynamical features - the expansion of erosion, soil moisture. Attempts to study land degradation processes and the necessity of degradation prediction have resulted in the creation of erosion models. The necessary information to modeling can be often derived from satellite images. The vegetative cover is a major factor of soil erosion.

Forest Mapping

One of the basic applications is forest cover typing and species identification. Forest cover typing will consist of reconnaissance mapping over a large area, while species inventories are highly detailed measurements of stand contents and characteristics. Using remote sensing data we can identify and delineate different forest types that would be difficult and time consuming using traditional ground surveys. Data is available at various scales and resolutions to satisfy local or regional demands. Requirements for reconnaissance mapping depend on the scale of study. For mapping differences in forest cover are required.

- Multi-spectral images, a very high resolution data is required to get detailed species identification.
- Multi-temporal images datasets contribute phenology information of seasonal changes of various species.
- Stereo photos help in the delineation and assessment of density, tree height and species.
- Hyper-spectral imagery will be used to generate signatures of vegetation species and certain stresses on trees. Hyper-spectral data offers a unique view of the forest cover, available only through remote sensing technology.
- RADAR is more useful for applications in the humid tropics because its all-weather imaging capability is valuable for monitoring forest.
- LIDAR data allows the 3-dimensional structure of the forest. The multiple return systems are capable of detecting the elevation of land and objects on it. The LIDAR data help estimate a tree height, a crown area and number of trees per unit area.

Clear Cut Mapping and Deforestation

The most important global problem is deforestation. There are many implications of it: in industrialized parts of world, pollution has damaged a large percentage of forested land, in tropical countries, valuable rain forest is being destroyed in an effort to clear potentially valuable agricultural and pasture land. The loss of forests increases soil erosion, river siltation and deposition, affecting the environment.

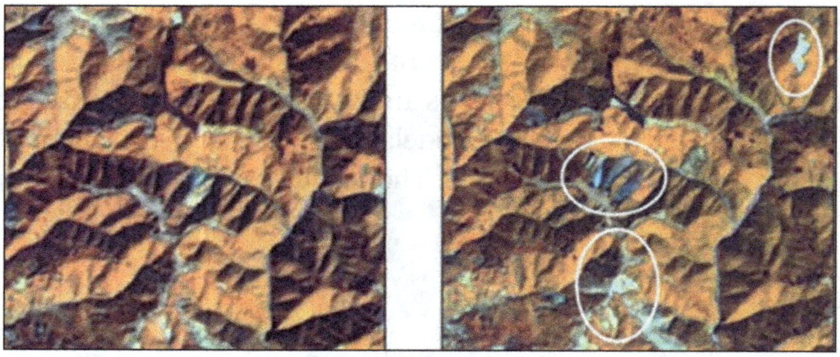

Forest cuts monitoring (LANDSAT TM 1992 and 2000).

Land Cover

This mapping is one of the most important and typical applications of remote sensing data. Land cover corresponds to the physical condition of the ground surface, for example, forest, grassland, concrete pavement etc., while land use reflects human activities like the use of the land, for example, industrial zones, residential zones, agricultural fields etc Initially the land cover classification system should be established, which is typically defined as levels and classes. The level and class should be designed in consideration of the purpose of use, the spatial and spectral resolution of the remote sensing data, user's request and so on.

Land cover change detection is necessary for updating land cover maps and the management of natural resources. The change is typically detected by comparison between two multi-date images, or sometimes between an old map and an updated remote sensing image.

Seasonal change: agricultural lands and deciduous forests change seasonally.

Annual change: land cover or land use changes, which are real changes, for example deforested areas or newly built towns.

Information on land cover and changing land cover patterns is directly useful for determining and implementing environment policy and will be used with other data to make complex assessments.

Corine Land Cover

This is the largest databases, providing information on the physical characteristics of the earth surface. Satellite images are utilized as the main source data to derive land cover information. The number of participating countries currently is 39. In the beginning of the project an information system on the state of the European environment was established, nomenclatures and methodologies were developed and agreed at European level. The aim of the European Union's CLC project was to map and get information on land cover at scale 1:100.000 for the whole Europe. In the beginning of 1990's the land cover of the contemporary European Union countries were mapped first time using satellite images, general classification criteria and quality standards. Classification was done by visual interpretation. In the 2000 the majority of the participating countries updated the CLC1990 database using the same method of interpretation of the satellite images. In addition to CLC1990 and CLC2000 also the land cover changes database was produced. CORINE Land Cover 2006 is the third European Land Cover inventory. Multi-temporal images were used to derive the minimum 5 ha land cover changes that occurred between 2000 and 2006. Some characteristics of the CORINE databases are as follows the table:

Some characteristics of the CORINE databases:

	CORINE 1:100.000	CORINE 1:50.000
Satellite images	Landsat TM	Landsat ETM+ SPOT PAN
Date of images	1990-1992	1990-1993
Number of categories	44	87
Minimal extension of the area (ha)	25	4
Minimal width of the area (m)	100	50

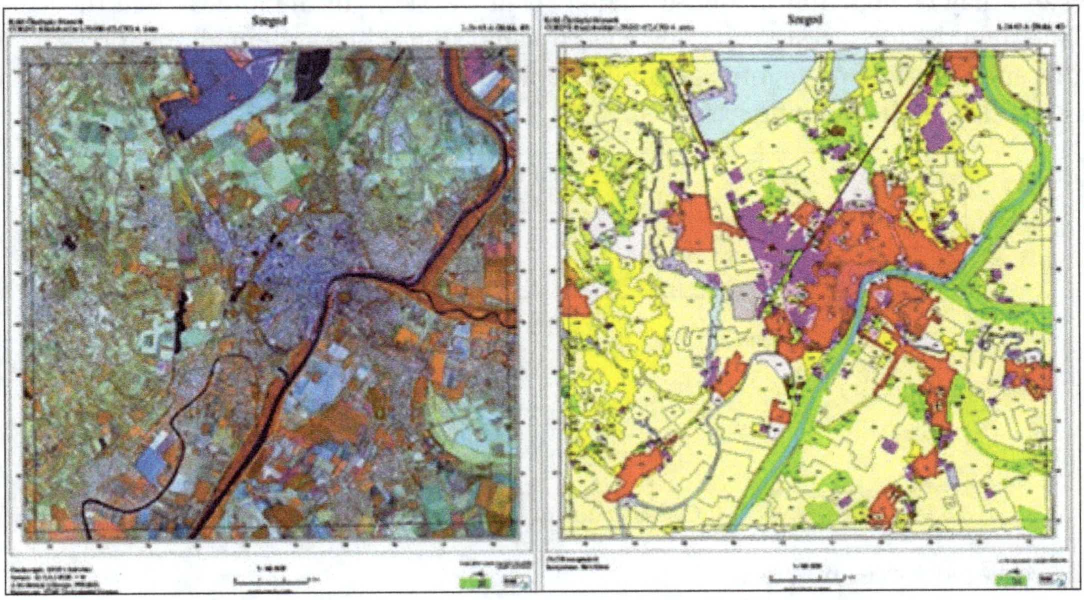

Spot satellite image as input data and the result of its interpretation.

Remote sensing data has shown tremendous potential for applications in various field for example in land use mapping and detection, geologic mapping, water resource applications, urban and regional planning, environment inventory, wetland mapping, natural disaster assessment or archaeological applications and other.

4

Forces and Equilibrium

4.1 Graphical and Analytical Treatment of Concurrent and Non-concurrent Co-planner Forces

Coplanar forces means the forces in a plane. When several forces act on a body then they are called as force system or a system of forces. A system in which all the forces lie in the same plane it is known as coplanar force system.

If the forces are having common line of action then they are known as collinear whereas if the forces intersect at a common point then they are known as concurrent. A force system may be coplanar or non-coplanar. If in a system all the forces lie in the same plane then the force system is known as coplanar. But if in a system all the forces lie in different planes then the force system is known as non-coplanar. Hence a force system is classified as shown in figure:

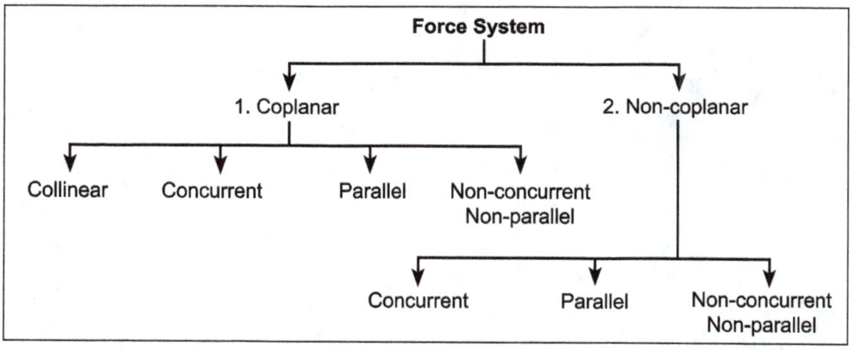

Classification of force system.

The coplanar forces may be:

Collinear

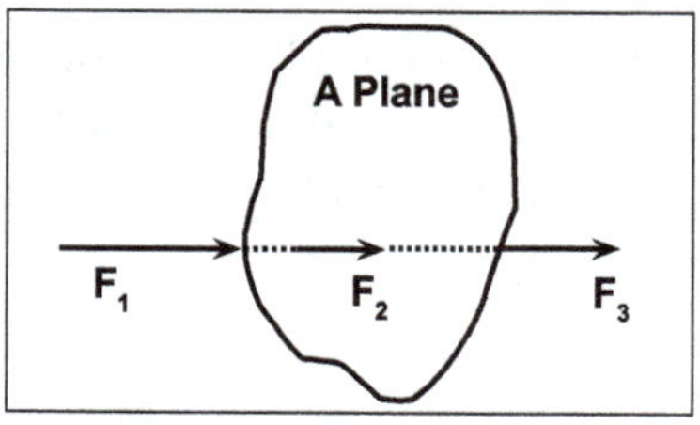

Coplanar collinear system.

A three forces F_1, F_2 and F_3 acting in a plane. These three forces are in the same line i.e., these three forces are having a common line of action. This system of forces is known as coplanar collinear force system. Hence in coplanar collinear system of forces all the forces act in the same plane and have a common line of action as shown in above figure.

Concurrent

A three forces F_1, F_2 and F_3 acting in a plane and these forces intersect or meet at a common point O. This system of forces is known as coplanar concurrent force system. Hence in coplanar concurrent system of forces all the forces act in the same plane and they intersect at a common point as shown in figure:

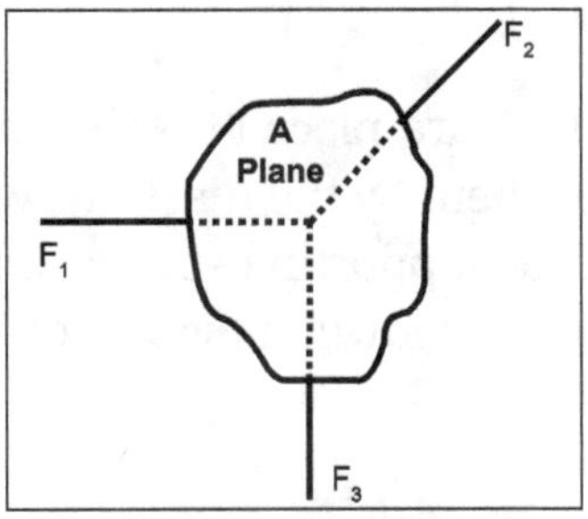

Coplanar concurrent force system.

Parallel

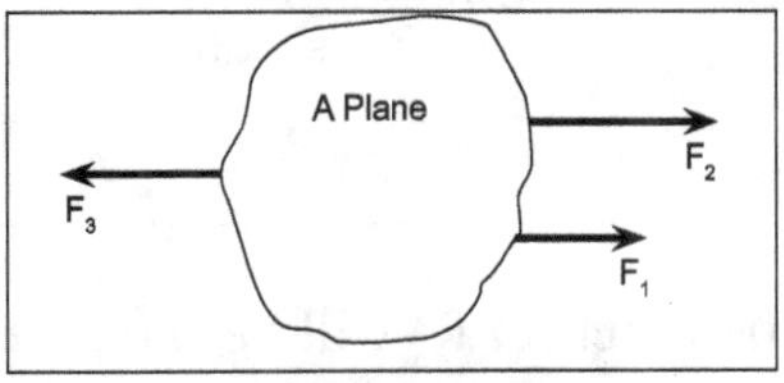

Parallel force system.

Figure shows three forces F_1, F_2 and F_3 acting in a plane and these forces are parallel. This system of forces is known as coplanar parallel force system. Hence in coplanar parallel system of forces all the forces act in the same plane and are parallel.

Coplanar Non-concurrent Non-parallel

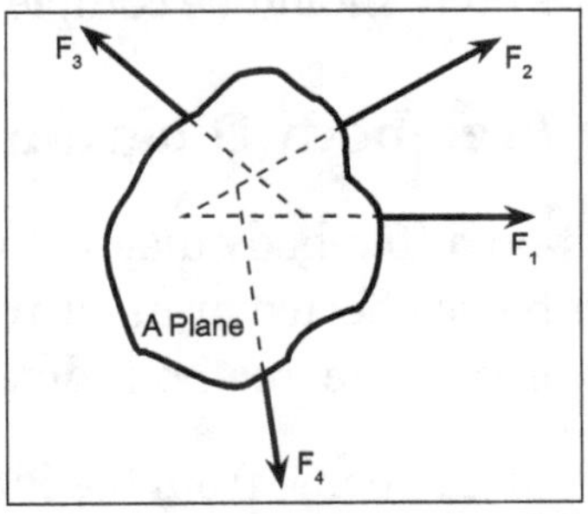

Coplanar Non-concurrent Non-parallel system.

Figure shows four forces F_1, F_2, F_3 and F_4 acting in a plane. The lines of action of these forces lie in the same plane but they are neither parallel nor meet or intersect at a common point. This system of forces is known as coplanar non-concurrent non-parallel force system. Hence in coplanar non-concurrent non-parallel system of forces all the forces act in the same plane but the forces are neither parallel nor meet at a common point. This force system is also known as general system of forces.

4.2 Free Diagram, Force Diagram and Bow's Notations

4.2.1 Free Body Diagram

The equilibrium of the bodies that are placed on the support can be considered if we remove the supports and replace them by the reactions which they exert on the body. In below Figure (a), if we remove the supporting surface and replace it by the reaction R_A that the surface exerts on the balls as shown in Figure (c), we shall get free-body diagram.

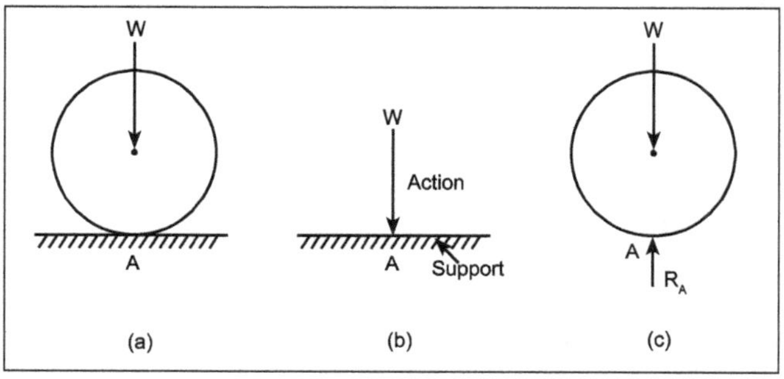

The point of application of the reaction R_A will be the point of contact A and from the law of equilibrium of two forces, we conclude that the reaction A must be vertical and equal to the weight W.

Consider a Figure (c), in which the ball is completely isolated from its support and all forces acting on the ball are shown by vectors is termed as free-body diagram. Therefore to draw the free-body diagram of a body we must remove all the supports such as wall, floor, hinge or any other body and replace those by the reactions with these support exert on the body where the body should be completely isolated.

Forces and Couples on a Free-body Diagram

Each force or the couple we put on a free-body diagram represents a model of how the body in the free body diagram is being effected by its surroundings. By selecting the forces and couples that are to be applied on the free-body diagram should follow these steps:

- First identify all the forces that come from the interaction of one body with another. Knowing the each way in which a support restricts the free motion of the body, a force or a moment must be applied to the body to impose the restriction on the motion.

- Then apply the weight of the body to its center of gravity.
- Know the strings and cables that can only pull on an object.
- Know the internal loads cancel out and should not be put on the free-body diagram.
- Finally, if we have selected the direction of forces or couples of interaction on one body, then Newton's 3rd law states that we must apply the forces or couples in the opposite direction on the other body.

The Graphical Method:

- Draw a load line for the applied loads and reactions to scale. Starting from the left-hand support.
- By using Bow's Notation as mentioned below, construct a force diagram, one joint at a time, in drawing each line parallel to the direction of the member in the framework.
- Then, load values that can now be scaled from the diagram.
- To determine the type of load in a member that can either tensile or compressive, should travel clockwise round a joint and note the force direction at the joint.
- Finally construct a table.

Bow's Notation

Bows notation is a method of lettering the cells and outside spaces which is formed by the directions of the stresses in and loads on a framed structure. Therefore these stresses and loads which can be traced by similar letters in the reciprocal diagram. It is also said to be a graphical method of representing the coplanar forces and stresses using alphabetical letters in the solution of stresses or in determining the resultant of a system of concurrent forces.

Bow's Notation is a convention for labelling the various zones in a diagram of a pin-jointed frame. It suggests the following:

- Letter the spaces between the external applied loads and reactions.
- Then number the spaces between the internal members.
- Always start with the letter 'A' between the reactions and work round the frame in a clockwise direction.

- Then start with the number 1 in the first left-hand space inside the framework.

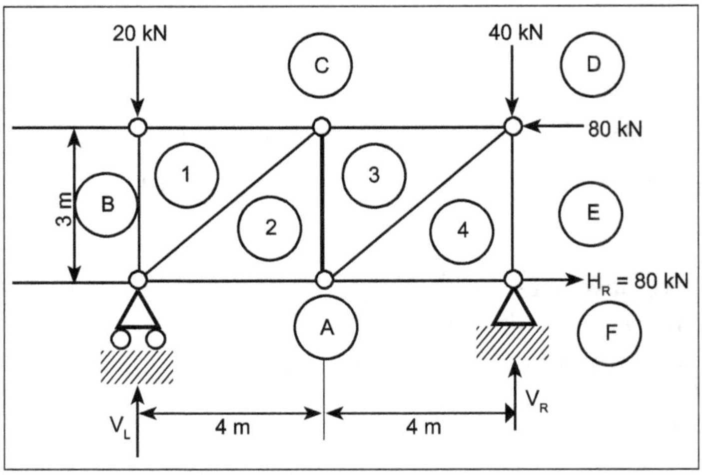

Application of Bow's Notation (Frame Diagram).

If the frame is labelled in accordance with Bow's Notation, it will appear. Check that the boundaries between the external zones are defined by the positions of the lines of the external forces and the reactions and members of the framework define the frontiers between the internal zones. Along the problem, we will be constructing a diagram termed as force diagram on a blank piece of graph paper. By doing this we will continually be referring back to the diagram, which is termed to be the frame diagram.

4.3 Application of Equilibrium Concepts

The concept of equilibrium describes a body which is stationary or which is moving with a constant velocity. Anybody under such a state is acted by the balanced forces and balanced couples. There is no unbalanced force or unbalanced couple acting on it. It is usually used in the analysis of a body which is stationary, or is said to be in the state of static equilibrium. It is the mostly used in engineering analysis.

Any basic concept when dealing with forces is said to be the idea of equilibrium or balance. An object can be acted on by several forces at the same time. A force is known as a vector quantity i.e., it has both a magnitude (size) and a direction associated with it. while the size and direction of the forces acting on an object should be exactly balanced, then there will be nonet force acting on the object. Therefore, the object is said to be in equilibrium. Since there is no net force acting on an object in equilibrium, then considering Newton's first law of motion, an object at rest will stay at rest, and an object in motion will stay in motion.

For example, the two forces acting on an object.

On the slide, we show a blue ball that is being pushed by two forces, the labeled Force

1 F1 and Force 2 F2 where the forces are vector quantities and direction is important. Two forces with the same magnitude but different directions are not equal forces. Thus:

$$F1 = -F2$$

Whereas for the coordinate system shown with the letter X below the ball. If we sum up the forces acting on the ball, we obtain the force equation on the left:

$$F1 + F2 = F\ net = 0$$

Here F net is the net force acting on the ball. Since the net force is equal to zero, the forces in example are acting in equilibrium.

Thus, there is no net force acting on the ball in example. Since the ball is initially at rest where velocity equals zero, the ball will remain at rest according to Newton's first law of motion. If the ball was travelling with the same uniform velocity, it might continue travelling at the same velocity. We have increased the magnitude of Force1 so that it is much greater than Force2. The forces are no longer in equilibrium. The force equation remains the same, but the net force is not equal to zero. The magnitude of the net force is given by:

$$F1 > -F2$$

$$F1 + F2 = F\ net$$

$$|F\ net| = |F1| - |F2|$$

Where the "| |" symbols indicate the magnitude of the quantity which is included between the ends. Therefore the direction of the net force would be in the positive X direction because F1 is greater than F2. Whereas according to Newton's second law of motion, the ball would begin to accelerate to the right. Because there is a net force in example, the forces are not in equilibrium.

Equilibrium

Force Equation:
$$F1 = -F2$$
$$F1 + F2 = Fnet = 0$$

No net external force Object at rest - stays at rest Newton's 1st Law of Motion

Non-equilibrium

Force Equation:
$$F1 > -F2$$
$$F1 - F2 = F\ net$$

Net external force Object accelerates Newton's 2nd Law of Motion.

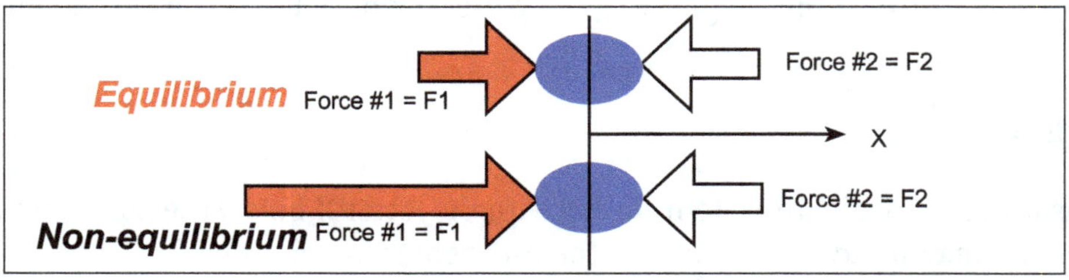

4.4 Analysis of Plane Trusses: Method of Joints and Method of Sections

A truss is one of the important types of engineering structures which gives a practical and economical solution for many engineering constructions, especially in the design of bridges and buildings that demand large spans.

The basic element of a plane truss is the triangle. Three bars jointed by pins at their ends constitute a rigid frame. The structure can be extended by adding each time two additional bars through a joint to form a rigid, i.e. non-collapsible structure.

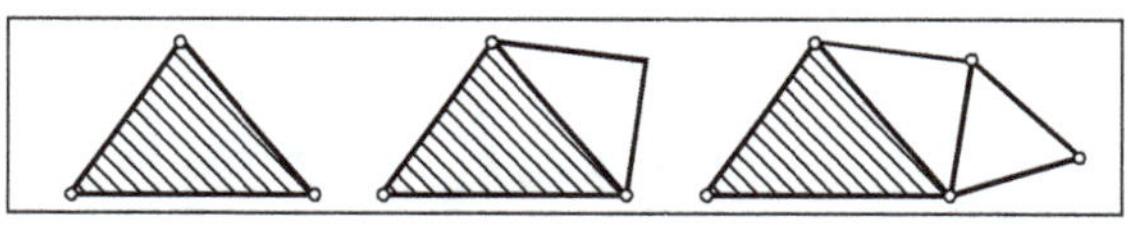

Plane trusses like on the sides of a bridge and space trusses like the TV towers.

<table>
<tr><td>Trusses</td><td>Frames</td></tr>
<tr><td>Members are subjected to tension/compression only.</td><td>At least one member is subjected to bending, shear or torsion.</td></tr>
<tr><td>Composed of rods.</td><td>Composed of bars.</td></tr>
<tr><td>Usually more simple to calculate, more weight effective.</td><td>Usually more simple to manufacture, and to place in tight layout.</td></tr>
<tr><td>This is a closed geometrical configuration with members jointed at ends only. Members are straight.

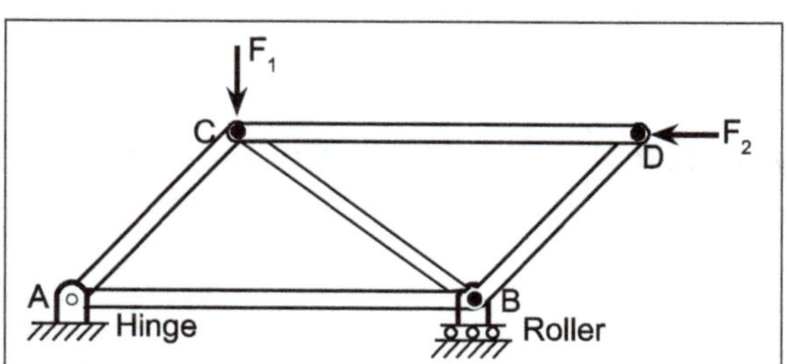
</td><td>It can be any configuration with members continuous beyond joints also. There can be bent members in a frame.

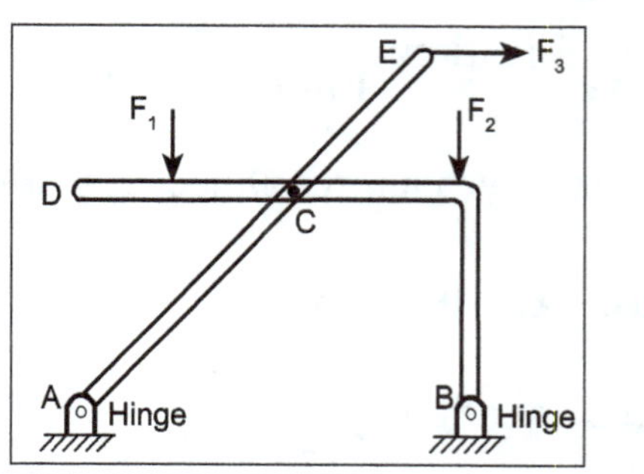
</td></tr>
</table>

Forces and loads are applied at joints only which makes the members to be two force. (There are 'axial forces' only).	Forces are applied anywhere on the member or pulley. Hence the members are typically 'multiforce'. (Sometimes two forces member may exist)
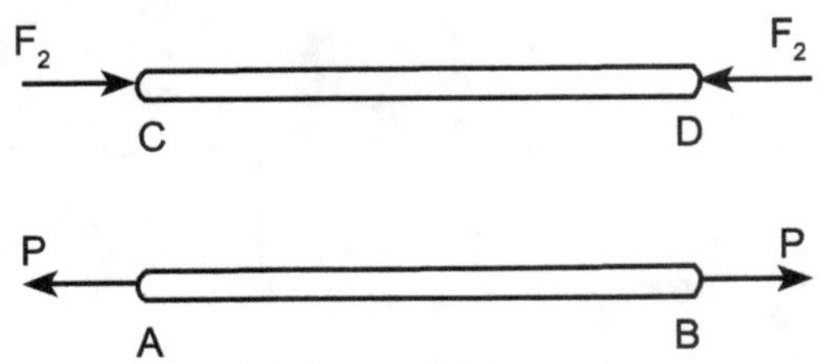 Two force members in truss	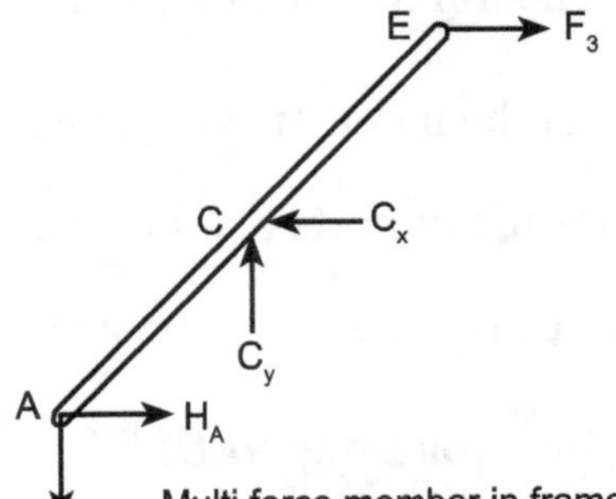 Multi force member in frame
Trusses are perfect, with a relationship between number of members and number of joints.	No such relationship exists. Number of joints and number of members vary from frame to frame.

The different types of frame are:

- Perfect frame
- Imperfect frame

Perfect Frame

If a frame is composed of such members, which are just sufficient to keep the frame in equilibrium, when the frame is supporting the external load then the frame is known as perfect frame.

Imperfect Frame

A frame in which number of members and number of joints are not given by $n = 2j - 3$ is known as imperfect frame. This means that number of members in an imperfect frame will be either more or less than $(2j - 3)$.

Deficient Frame

If the number of member in a frame are less than $(2j - 3)$, then the frame is known as deficient frame.

Redundant frame

If the number of member in a frame are more than $(2j - 3)$, then the frame is known redundant frame.

The assumptions made in finding out the forces in a frame are:

- The frame is a perfect frame.
- The frame carries load at the joints.
- All the members are pin-joined.

Reactions of Supports of a Frame:

The frame are generally supported;

- On a roller support.
- On a hinged support.

Method of Joints

Method of Joints Applied to Trusses Carrying Horizontal Loads:

If a truss carries horizontal loads either with or without vertical loads attached at one end supported on roller at the other end, the support reaction at the roller support end will be normal. The support reaction at the attached end will consist of,

- Horizontal reaction.
- Vertical reaction.

Forces in a Member by Method of Joints

While determining the forces in a member by methods of joints, the joint should be selected in such a way that at any time there are only two members in which the forces ıre unknown.

thod of Sections

ethod of sections is another method to determine the forces in members of a ·ucture. In order to find the unknown forces in the method of sections, the sec- ·he truss structure must be isolated.

·e on the entire isolated section must be zero because the isolated section ·e. This method is compartively faster because instead of moving from ·ith the method of joints the members of interest can be immediately

as

Problems

1. Here we will discuss a problem based on method of joints.

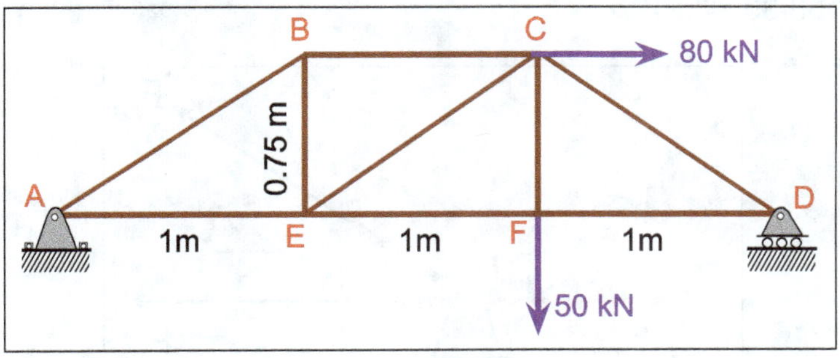

Solution:

Given:

Load acting in c = 80KN

Load acting in F = 50KN

BE = 0.75

AE = 1m

EF = 1m

FD = 1m

Conditions:

$$\Sigma M = 0, \ \Sigma H = 0, \ \Sigma V = 0$$

$$\sum M_D = 0$$

$$3A_V + 50(1) = 80(0.75)$$

$$A_V = 3.33\text{kN}$$

$$\sum F_H = 0$$

$$A_H = 80\text{kN}$$

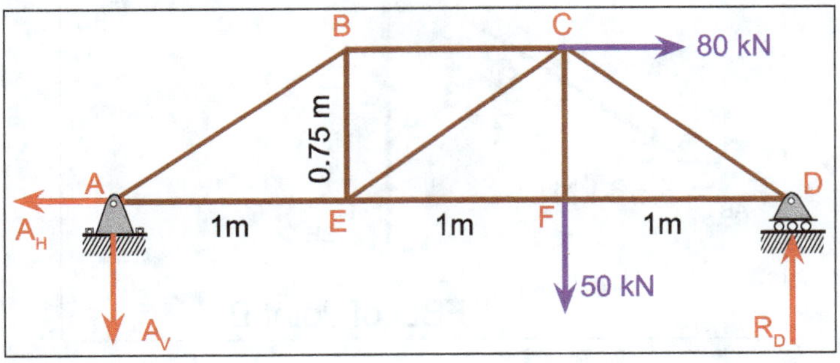

$$\sum M_A = 0$$

$$3R_D = 50(2) + 80(0.75)$$

$$R_D = 53.33\,\text{kN}$$

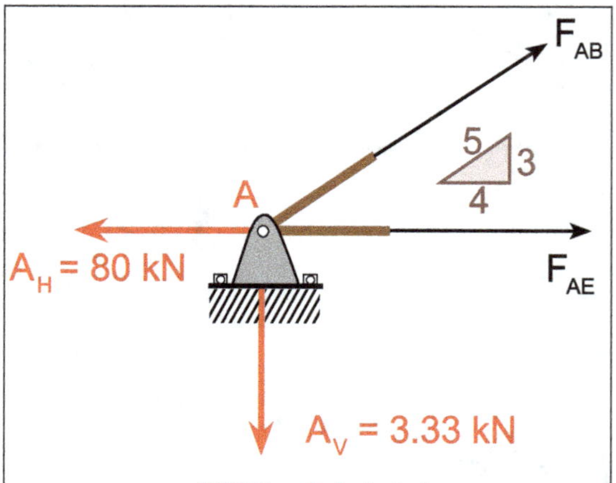

FBD of Joint A

At joint A

$$\sum F_V = 0$$

$$\frac{3}{5}F_{AB} = 3.33$$

$$F_{AB} = 5.56\,\text{kN tension}$$

$$\sum F_H = 0$$

$$F_{AE} + \frac{4}{5}F_{AB} = 80$$

$$F_{AE} + \frac{4}{5}(5.56) = 80$$

$$F_{AE} = 75.56\ \text{kN tension}$$

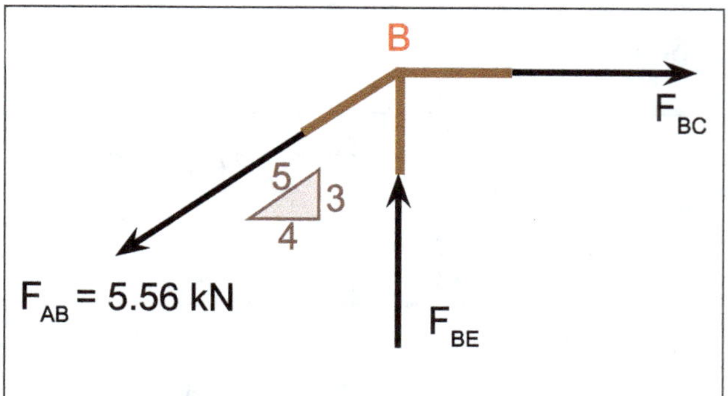

FBD of Joint B

At joint B

$$\sum F_H = 0$$

$$F_{BC} = \frac{4}{5} F_{AB}$$

$$F_{BC} = (5.56)$$

$$F_{BC} = 4.45 \text{ kN tension}$$

$$\sum F_V = 0$$

$$F_{BE} = \frac{3}{5} F_{AB}$$

$$F_{BE} = \frac{3}{5}(5.56)$$

$$F_{BE} = 3.34 \text{ kN compression}$$

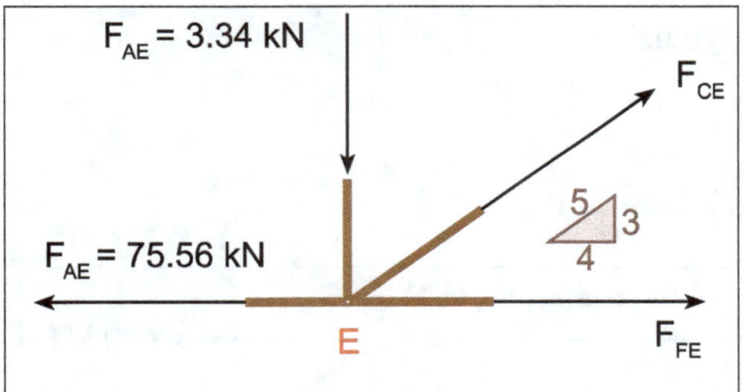

FBD of Joint E

At joint E

$$\sum F_V = 0$$

$$\frac{3}{5} F_{CE} = F_{AE}$$

$$\frac{3}{5} F_{CE} = 3.34$$

$$F_{CE} = 5.57 \text{ kN tension}$$

$$\sum F_H = 0$$

$$F_{FE} + \frac{4}{5} F_{CE} = F_{AE}$$

$$F_{FE} + \frac{4}{5}(5.57) = 75.56$$

$$F_{FE} = 71.11 \text{ kN tension}$$

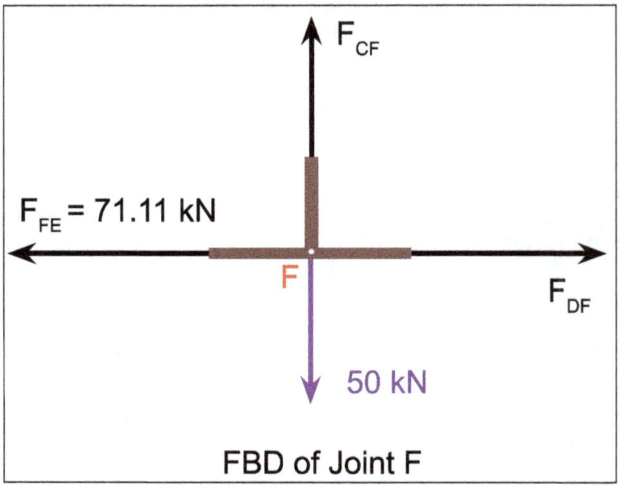

FBD of Joint F

At joint F

$\sum F_V = 0$

$F_{CE} = 50\,\text{kN tension}$

$\sum F_H = 0$

$F_{DF} = F_{FE}$

$F_{DF} = 71.11\,\text{kN tension}$

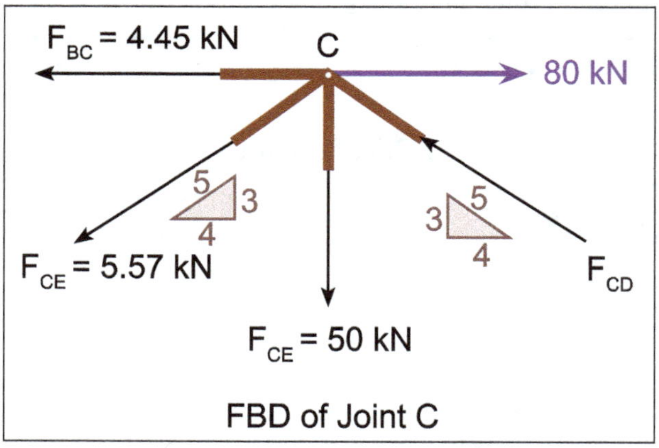

FBD of Joint C

At joint C

$\sum F_H = 0$

$$\frac{4}{5}F_{CD} + \frac{4}{5}F_{CE} + F_{BC} = 80$$

$$\frac{4}{5}F_{CD} + \frac{4}{5}(5.57) + 4.45 = 80$$

$F_{CD} = 88.87\,\text{kN compression}$

$$\sum F_V = 0$$

$$\frac{3}{5}F_{CD} = \frac{3}{5}F_{CE} + F_{CF}$$

$$\frac{3}{5}(88.87) = \frac{3}{5}(5.57) = 50$$

$$53.3 = 53.3$$

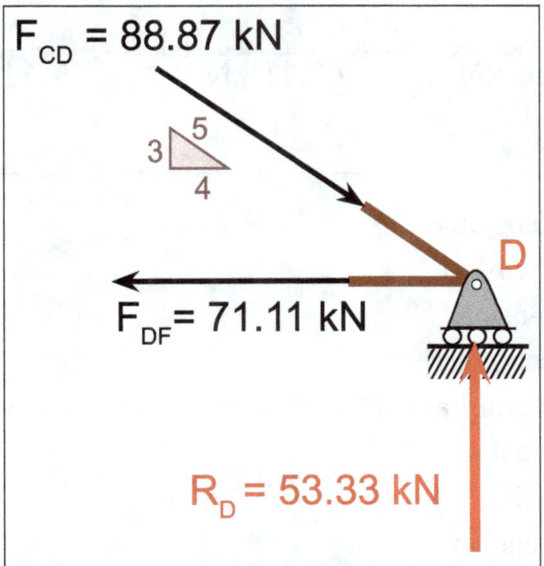

FBD of Joint D

At joint D

$$\sum F_H = 0$$

$$\frac{4}{5}F_{CD} = F_{DF}$$

$$\frac{4}{5}(88.87) = 71.11$$

$$71.1 = 71.1$$

$$\sum F_V = 0$$

$$R_D = \frac{3}{5}F_{CD}$$

$$53.33 = \frac{3}{5}(88.87)$$

$$53.3 = 53.3$$

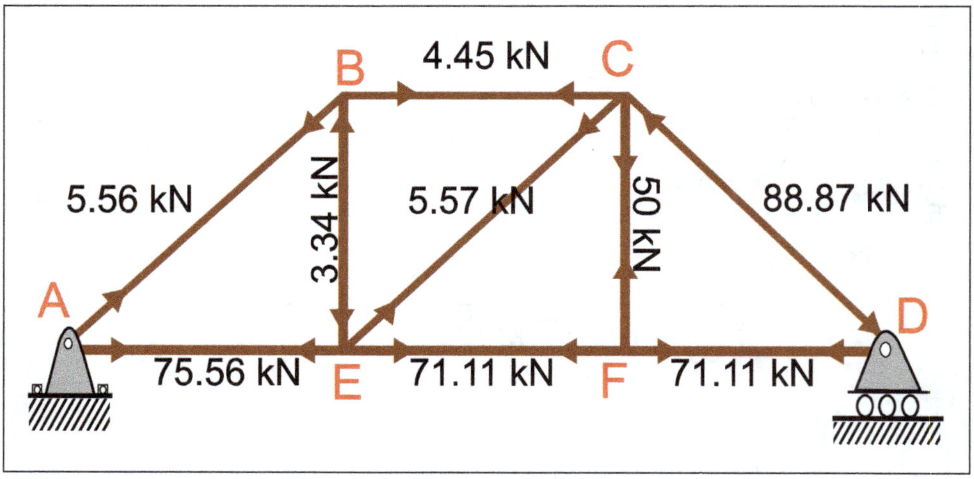

$F_{AB} = 5.56$ kN tension
$F_{AE} = 75.56$ kN tension
$F_{BC} = 4.45$ kN tension
$F_{BE} = 3.34$ kN compression
$F_{CD} = 88.87$ kN compression
$F_{CE} = 5.57$ kN tension
$F_{CF} = 50$ kN tension
$F_{DF} = 71.11$ kN tension
$F_{FE} = 71.11$ kN tension

2. Using the method of joints, let us determine the force in each member of the truss.

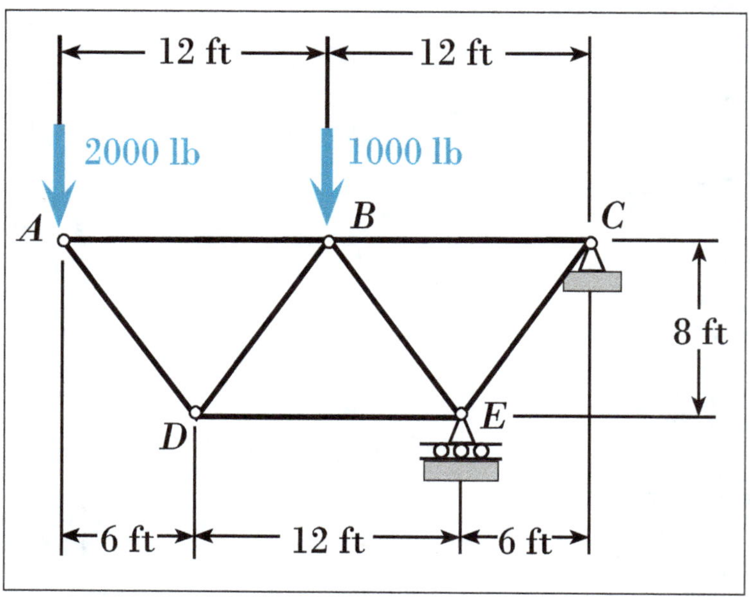

Solution:

Given:

EC = 8ft

DE = 12ft

Load at A = 2000lb

Load at B = 1000lb

Conditions:

$$\sum M = 0,\ \sum H = 0,\ \sum V = 0$$

Draw the free body diagram of the truss and solve for the equations.

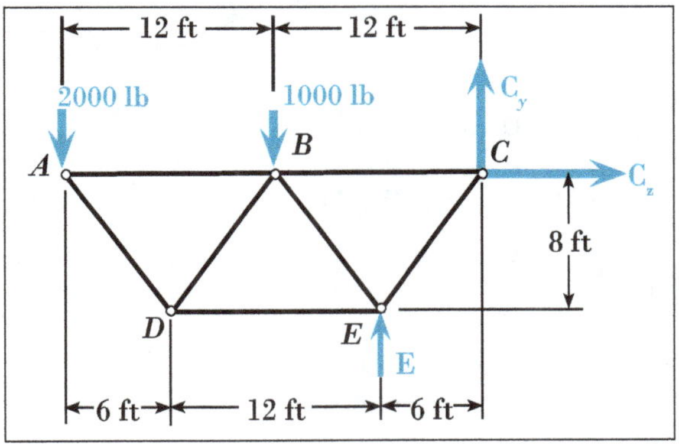

$$\sum F_X = 0 = C_X$$

$$C_X = 0 \text{ lb}$$

$$\sum F_Y = 0 = -2000 \text{ lb} - 1000 \text{ lb} + E + C_y$$

$$\Rightarrow E + C_Y = 3000 \text{ lb}$$

Solve the moment about C:

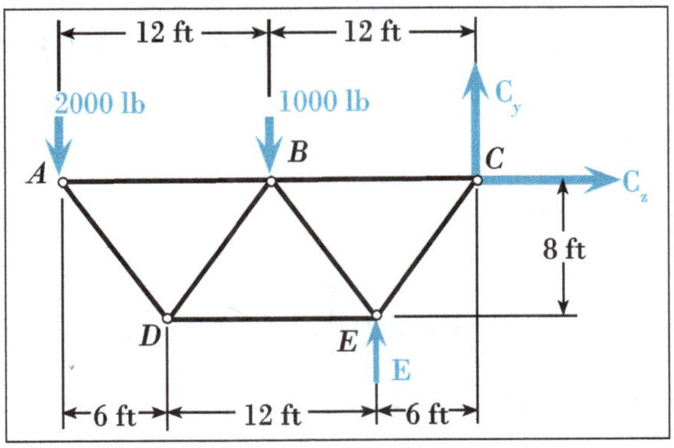

$$\sum M_C = 0 = 2000\,lb(24\,ft) + 1000\,lb(12\,ft) - E(6\,ft)$$
$$\Rightarrow E = 10000\ lb$$
$$\Rightarrow C_y = 3000\ lb - 10000\ lb = -7000\ lb$$

Look at joint A:

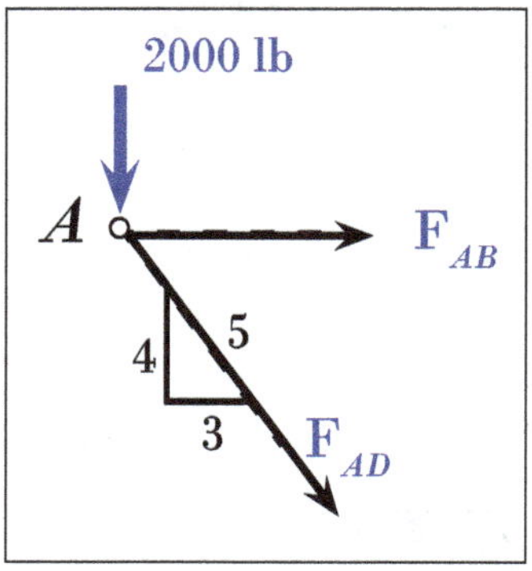

$$\sum F_Y = 0 = -\frac{4}{5}F_{AD} - 2000\ lb$$
$$F_{AD} = -2000\ lb \Rightarrow F_{AD} = 2500\ lb(C)$$
$$\sum F_X = 0 = \frac{3}{5}F_{AD} + F_{AB} = \frac{3}{5}(-2500\ lb) + F_{AB}$$
$$F_{AB} = 1500\ lb \Rightarrow F_{AB} = 1500\ lb(T)$$

Look at joint D:

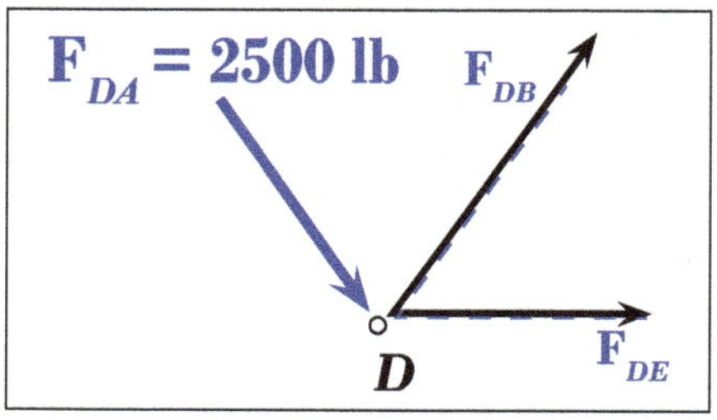

$$\sum F_Y = 0 = \frac{4}{5}F_{AD} + \frac{4}{5}F_{DB} = \frac{4}{5}(-2500\ lb) + \frac{4}{5}F_{DB}$$
$$F_{DB} = 2500\ lb \Rightarrow F_{DB} = 2500\ lb(T)$$
$$\sum F_X = 0 = -\frac{3}{5}F_{AD} + \frac{3}{5}F_{DB} + F_{DE}$$
$$= \frac{-3}{5}(-2500\ lb) + \frac{3}{5}(2500\ lb) + F_{DE}$$
$$F_{DE} = -3000\ lb\ Þ\ F_{DE} = 3000\ lb(C)$$

Look at joint B:

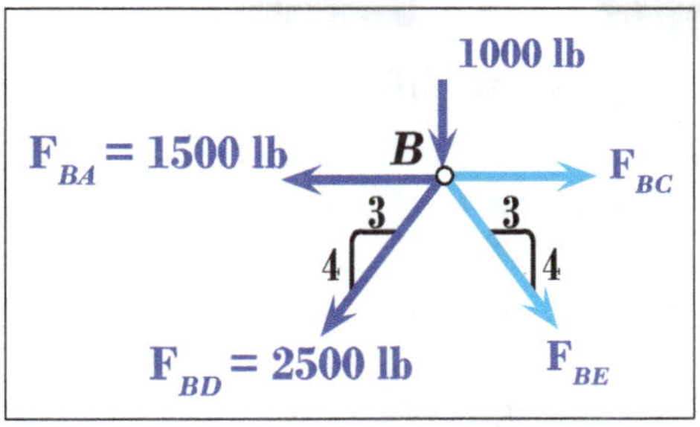

$$\sum F_Y = 0 = -\frac{4}{5}F_{BD} - \frac{4}{5}F_{BE} - 1000 \text{ lb}$$

$$= -\frac{4}{5}(2500 \text{ lb}) - \frac{4}{5}F_{DE} - 1000 \text{ lb}$$

$$F_{DE} = -3750 \text{ lb} \Rightarrow F_{DE} = 3750 \text{ lb (C)}$$

$$\sum F_X = 0 = -\frac{3}{5}F_{BD} - F_{BA} + \frac{3}{5}F_{BE} + F_{BC}$$

$$= \frac{-3}{5}(-2500 \text{ lb}) - 1500 \text{ lb} + \frac{3}{5}(-3750 \text{ lb}) + F_{BC}$$

$$F_{BC} = 5250 \text{ lb} \Rightarrow F_{DE} = 5250 \text{ lb(T)}$$

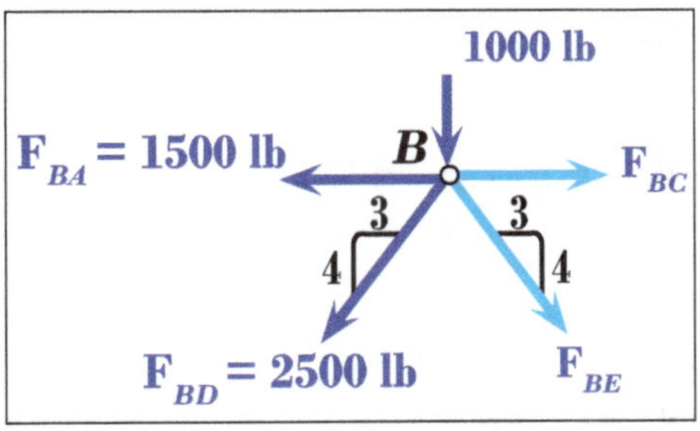

Look at joint E:

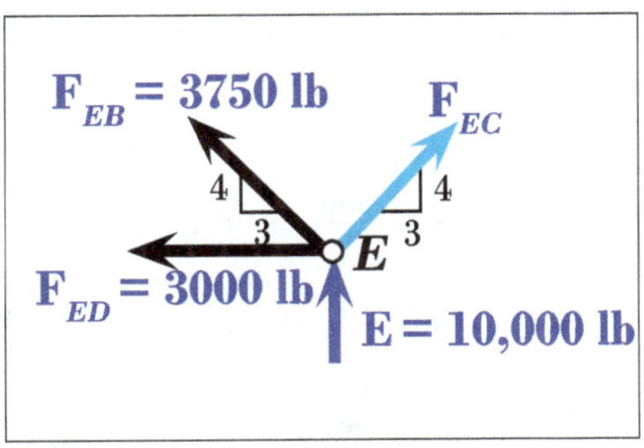

$$\sum F_Y = 0 = -\frac{4}{5}F_{EB} + \frac{4}{5}F_{EC} + 10000 \text{ lb}$$

$$= -\frac{4}{5}(-3750\text{ lb}) + \frac{4}{5}F_{DE} + 10000\text{ lb}$$

$$F_{EC} = -8750\text{ lb} \Rightarrow F_{EC} = 8750\text{ lb}(C)$$

$$\sum F_X = 0 = -\frac{3}{5}F_{EB} - F_{ED} + \frac{3}{5}F_{EC}$$

$$= -\frac{3}{5}(-3750\text{ lb}) - (-3000\text{ lb}) + \frac{3}{5}F_{EC}$$

$$F_{EC} = -8750\text{ lb} \Rightarrow F_{EC} = 8750\text{ lb}(C)$$

Look at joint C to check the solution:

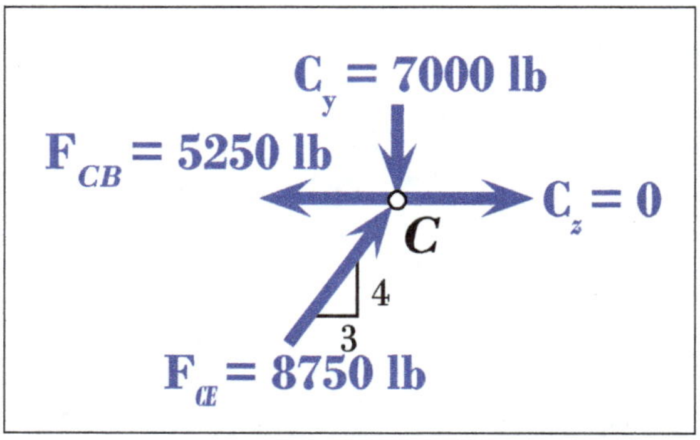

$$\sum F_Y = 0 = -\frac{4}{5}F_{CE} - 7000\text{ lb}$$

$$= -\frac{4}{5}(-8750\text{ lb}) - 7000\text{ lb} = 0\text{ OK!}$$

$$\sum F_X = 0 = -\frac{3}{5}F_{CE} - F_{CB} + C_x$$

$$= -\frac{3}{5}(-8750\text{ lb}) - (5250\text{ lb}) + 0 = 0$$

3. Let us determine the force in each member of the loaded truss by Method of Joints.

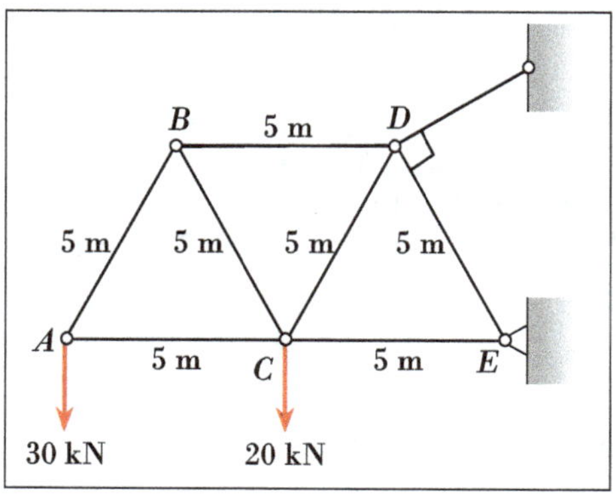

Solution:

Given:

Load at A = 30kN

Load at C = 20kN

AB = AC = BC = BD = CD = DE = CE = 5m

Conditions:

$\sum M = 0,\ \sum H = 0,\ \sum V = 0$

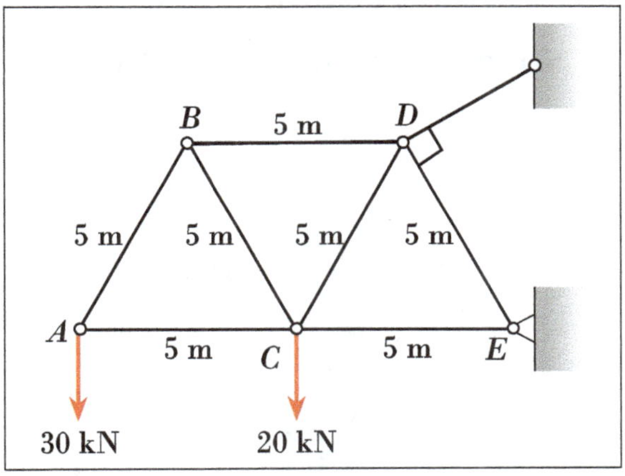

$[\sum M_E = 0] \quad 5T - 20(5) - 30(10) = 0 \quad T = 80\ \text{kN}$

$[\sum F_x = 0] \quad 80\cos 30° - E_x = 0 \quad E_x = 69.3\ \text{kN}$

$[\sum F_y = 0] \quad 80\sin 30° + E_y - 20 - 30 = 0 \quad E_y = 10\ \text{kN}$

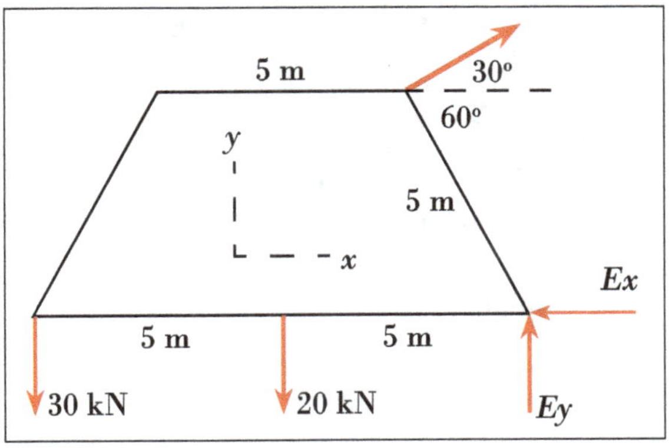

$[\sum F_y = 0] \quad 0.866AB - 30 = 0 \quad AB = 34.6\ \text{kN T}$

$[\sum F_x = 0] \quad AC - 0.5(34.6) = 0 \quad AC = 17.32\ \text{kN C}$

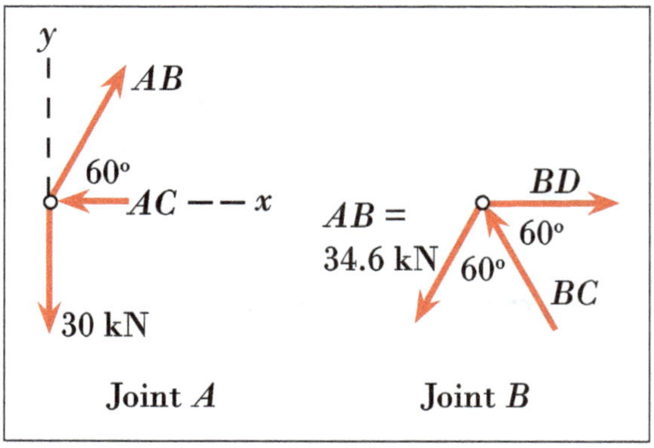

$$\left[\sum F_y = 0\right] \quad 0.866BC - 0.866(34.6) = 0 \quad BC = 34.6 \text{ kN C}$$

$$\left[\sum F_x = 0\right] \quad BD - 2(0.5)(34.6) = 0 \quad BD = 34.6 \text{ kN T}$$

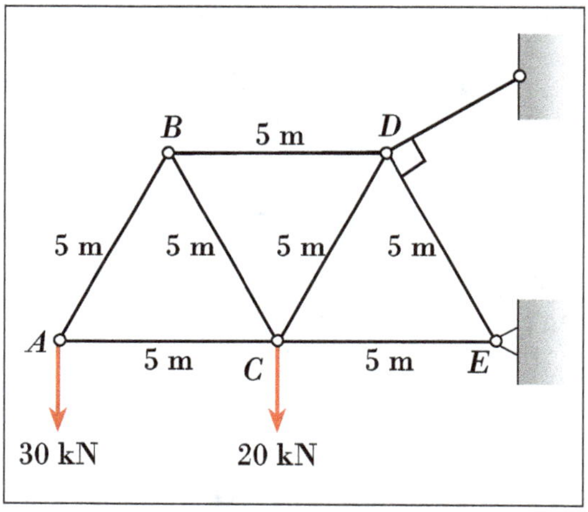

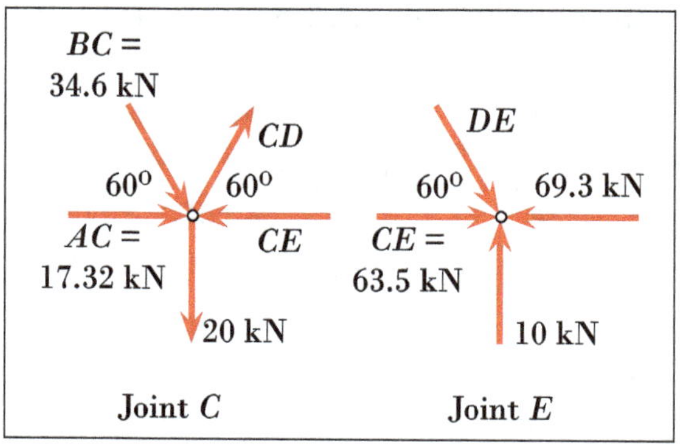

$$\left[\sum F_y = 0\right] \quad 0.866CD - 0.866(34.6) - 20 = 0$$

$$CD = 57.7 \text{ kN T}$$

$$\left[\sum F_x = 0\right] \quad CE - 17.32 - 0.5(34.6) - 0.5(57.7) = 0$$

$$CE = 63.5 \text{ kN C}$$

$\left[\sum F_y = 0\right]$ $0.866DE = 10$ $DE = 11.55$ kN C

and the equation $\sum F_x = 0$ checks.

4. Let us discuss a problem based on method of section.

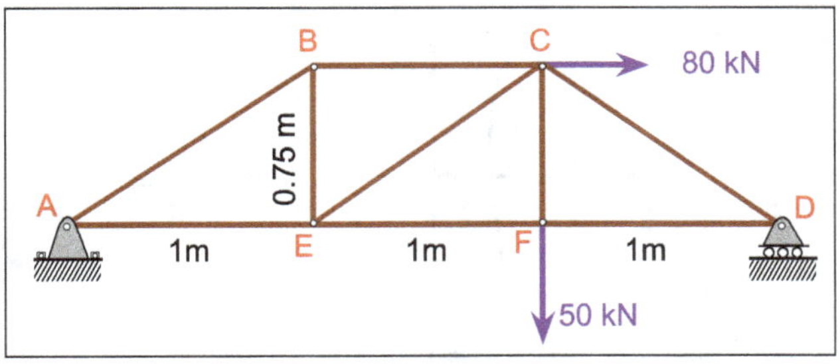

Solution:

Given:

Load at C = 80kN

Load at F = 50kN

AE = EF = FD = 1m

Conditions:

$\sum M = 0, \ \sum H = 0, \ \sum V = 0$

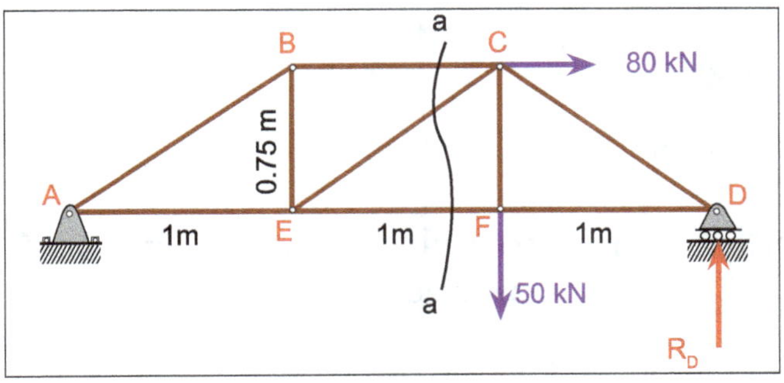

$$\sum M_A = 0$$
$$3R_D = 50(2) + 80(0.75)$$
$$R_D = 53.33 \text{ kN}$$

From the FBD of the section through a – a

$$\sum M_E = 0$$
$$0.75 \ F_{BC} + 2R_D = 0.75(80) + 1(50)$$

$$0.75\,F_{BC} + 2(53.33) = 60 + 50$$
$$F_{BC} = 4.45 \text{ kN tension}$$

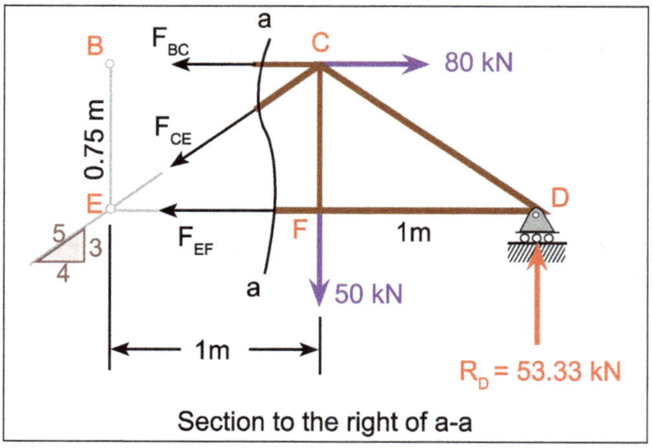

Section to the right of a-a

$$\sum M_C = 0$$
$$0.75F_{EF} = 1(R_D)$$
$$0.75F_{EF} = 53.33$$
$$F_{EF} = 71.11 \text{ kN tension}$$

$$\sum F_V = 0$$
$$\frac{3}{5}F_{CE} + 50 = R_D$$
$$\frac{3}{5}F_{CE} + 50 = 53.33$$
$$F_{CE} = 5.55 \text{ kN tension}$$

5. Use the line of action of the forces and take the moment about G it will remove the F_{GI} and F_{GH} and shift F_{FH} to the perpendicular of G.

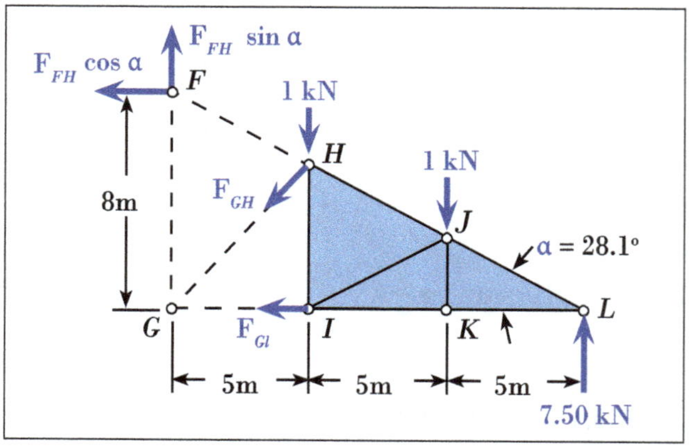

Solution:

Given:

Load at L = 7.50kN

Load at H = 1kN

Load at J = 1kN

FG = 8m

GI = IK = KL = 5m

Conditions:

$$\sum M=0,\ \sum H=0,\ \sum V=0$$

Take the moment at G:

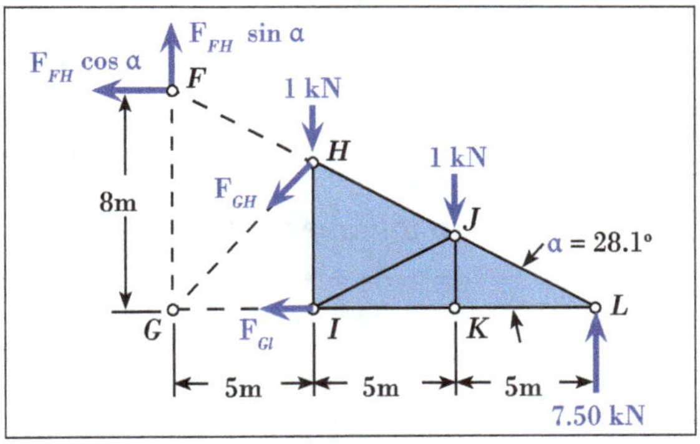

$$\sum M_G=-1\text{kN}(5\text{ m})-1\text{kN}(10\text{ m})+7.5\text{kN}(15\text{ m})$$
$$+F_{FH}\cos(28.1°)(8\text{ m})$$
$$F_{FH}=-13.82\text{ kN}\Rightarrow F_{FH}=13.82\text{ kN}(C)$$

Use the line of action of the forces and take the moment about L it will remove the F^{GI} and F^{FH} and shift F^{GH} to point G.

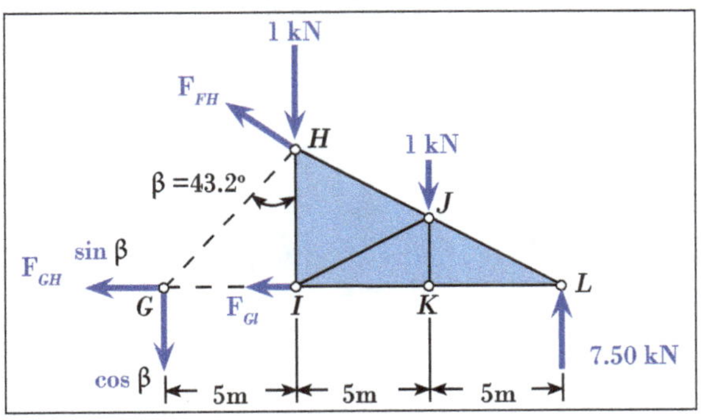

$$\beta = \tan^{-1}\left(\frac{-5\,m}{-5.333\,m}\right) = -133.2°$$

Take the moment at L:

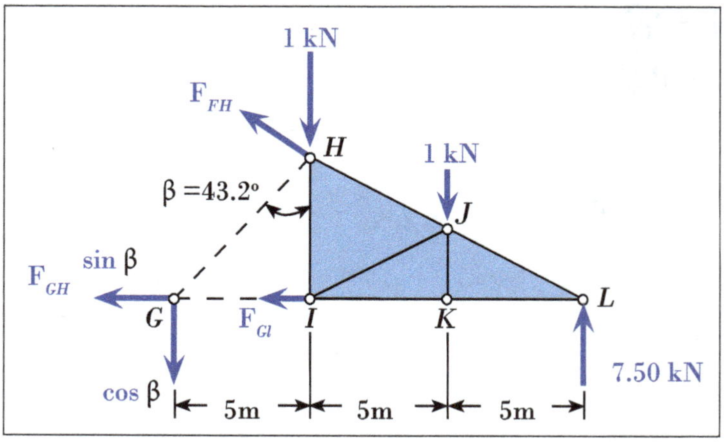

$$\sum M_L = 1\,kN(5\,m) + 1\,kN(10\,m) + F_{GH}\cos(43.2°)(15\,m)$$

$$F_{GH} = 1.372\,kN \Rightarrow F_{GH} = 1.372\,kN(C)$$

6. Let us find out the internal forces in members FH, GH, and GI.

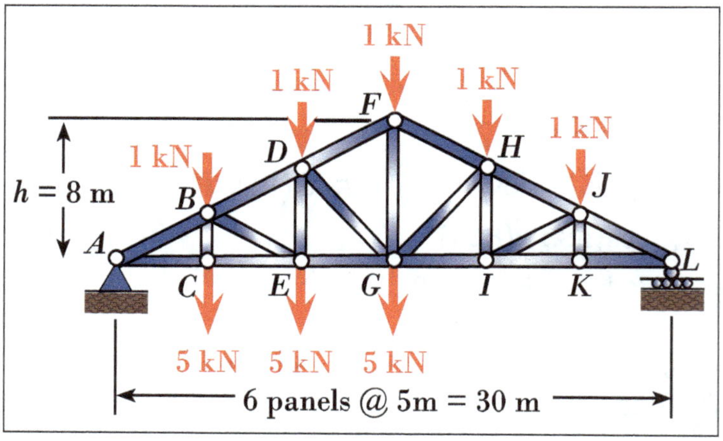

Solution:

Load = 1kN and 5kN

Length = 30m

Conditions:

$$\sum M = 0,\ \sum H = 0,\ \sum V = 0$$

The reactions are:

$$\sum M_A = 0 = -(5\text{ m})(6\text{kN}) - (10\text{m})(6\text{kN}) - (15\text{m})(6\text{kN})$$
$$-(20\text{m})(1\text{kN}) - (25\text{m})(1\text{kN}) + (30\text{ m})L$$
$$L = 7.5\text{ kN}\uparrow$$
$$\sum F_y = 0 = -20\text{kN} + L + A$$
$$A = 12.5\text{kN}\uparrow$$

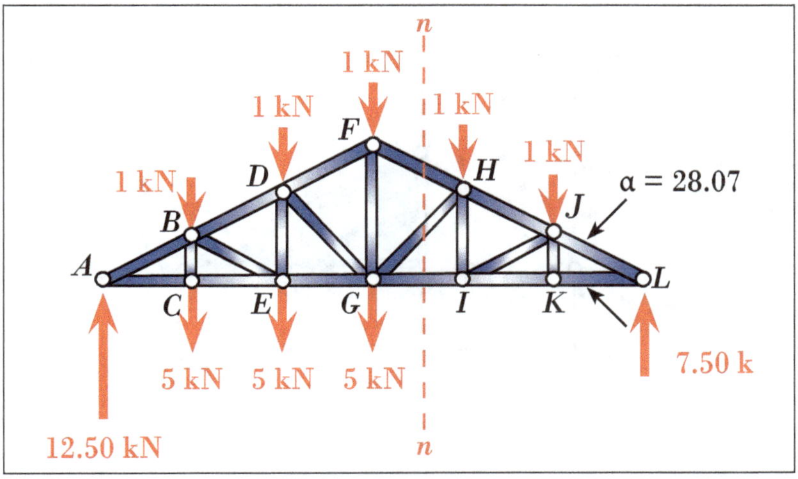

Pass a section through members FH, GH, and GI and take the right-hand section as a free body:

$$\tan\alpha = \frac{FG}{GL} = \frac{8\text{m}}{15\text{m}} = 0.5333 \quad \alpha = 28.07°$$

Pass a section through members FH, GH, and GI and take the right-hand section as a free body:

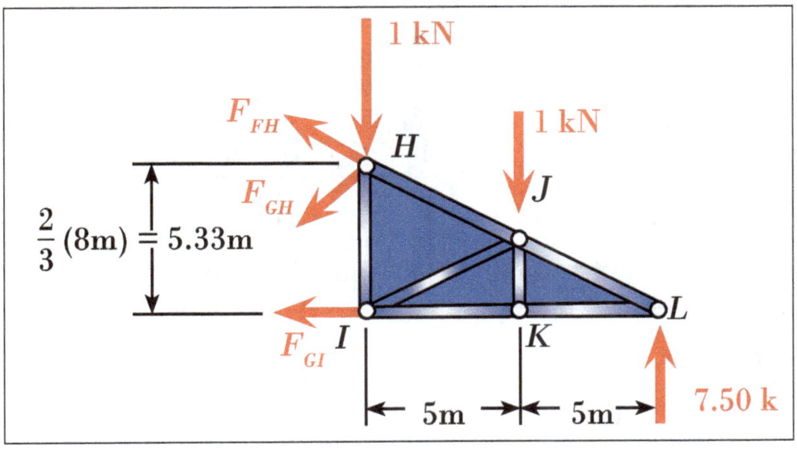

Apply the conditions for static equilibrium to determine the desired member forces:

$$\sum M_H = 0$$

$$(7.50\ \text{kN})(10\ \text{m}) - (1\text{kN})(5\ \text{m}) - F_{GI}(5.33\ \text{m}) = 0$$

$$F_{GI} = +13.13\ \text{kN}$$

$$F_{GI} = 13.13\ \text{kN T}$$

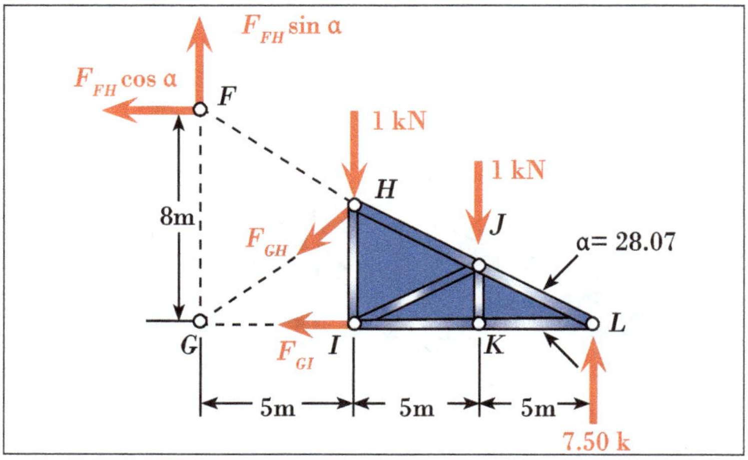

$$\sum M_G = 0$$

$$(7.5\ \text{kN})(15\ \text{m}) - (1\ \text{kN})(10\ \text{m}) - (1\ \text{kN})(5\ \text{m})$$

$$+(F_{FH} \cos \alpha)(8\ \text{m}) = 0$$

$$F_{FH} = -13.82\ \text{kN}$$

$$F_{FH} = -13.82\ \text{kN C}$$

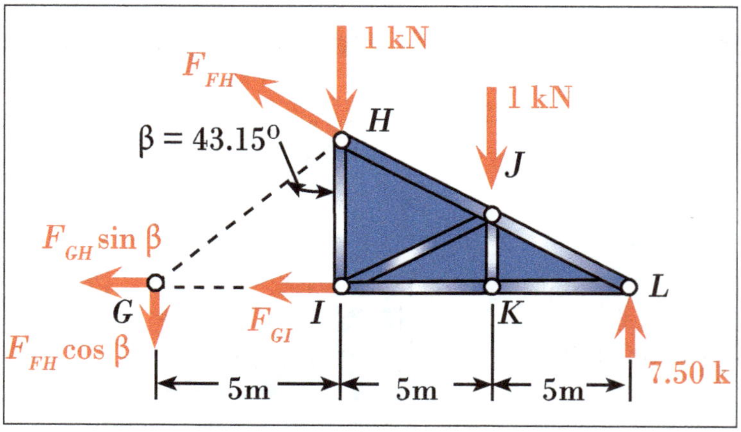

$$\tan \beta = \frac{GI}{HI} = \frac{5\text{m}}{\frac{2}{3}(8\text{m})} = 0.9375 \qquad \beta = 43.15°$$

$$\sum M_L = 0$$
$$(1\text{ kN})(10\text{ m}) + (1\text{ kN})(5\text{ m}) + (F_{GH}\cos\beta)(15\text{ m}) = 0$$
$$F_{GH} = -1.371\text{ kN}$$
$$F_{GH} = -1.371\text{ kN C}$$

4.5 Frictional Force in Equilibrium Problems

Friction may be defined as a force of resistance acting on a body which prevents or retards slipping of the body relative to a second body and surface with which it is in control.

1. Laws of Static Friction

- The force of friction always acts in a direction opposite to that in which the body tends to move if the force of friction would have been absent.
- The magnitude of the force of friction is exactly equal to the force which tends to move the body.
- The magnitude of the limiting friction bears a constant ratio to the normal reaction between the two surfaces mathematically such as:
    - F/R = Constant.
    - F= limiting force.
    - R= Normal force.
- The force of friction is independent of the area of contact between the two surfaces.
- The force of friction depends upon the roughness of the surfaces.

2. Laws of Kinetic or Dynamic Friction

- The force of friction always acts in a direction opposite to that in which the body is moving.
- The magnitude of kinetic friction bears a constant ratio to the normal reaction between the two surfaces. But the ratio is slightly less than that in case of limiting friction.
- For moderate speeds the force of friction remains constant. But it decreases slightly with the increase of speed.

Laws of Sliding Friction

The sliding of a solid body in contact with another solid body is always opposed by force of friction. The friction acts in the direction opposite to that of relative motion and it is tangential to the surface of two bodies at the point of contact.

The friction is a necessary in every machine because it involves wearing of machine component and consumes energy that transfers into heat. In some cases friction is desirable in case for functioning of a machine such as belt drives, friction clutches.

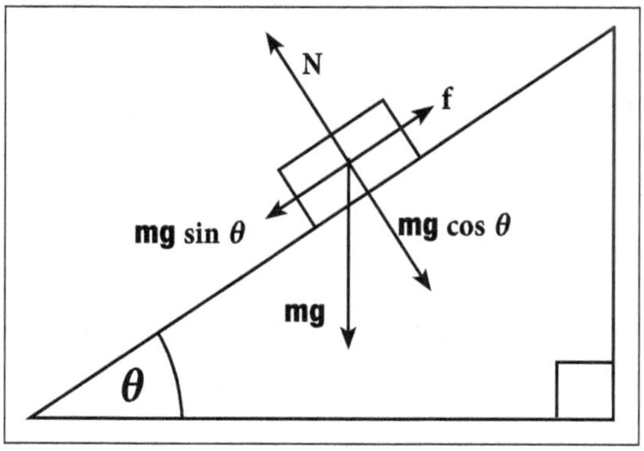

- Law 1: When two bodies are in contact with the direction of the forces of Friction on one of them at its point of contact is opposite to the the direction in which the point of contact tends to move relative to the other.
- Law 2: If the bodies are in equilibrium the force of friction is just sufficient to prevent motion and therefore it can be determined by applying the conditions of equilibrium of all the forces acting on the body. The amount of Friction that can be exerted between two surfaces is limited and if the forces acting on the body are made sufficiently great motion will occur. Hence we define the limiting friction as the friction which is exerted when equilibrium is on the point of being broken by one body sliding on another. The magnitude of limiting friction is given by the following three laws.
- Law 3: The ratio of the limiting friction to the normal reaction between two surfaces depends on the substances of which the surfaces are composed and not on the magnitude of the normal reaction.
- Law 4: The amount of limiting friction is independent of the area of contact between the two surfaces and the shape of the surfaces provided that the normal reaction is unaltered.
- Law 5: When motion takes place the direction of friction is opposite to the direction of relative motion and is independent of velocity. The magnitude of the force of friction is in a constant ratio to the normal reaction but this ratio may be slightly less than when the body is just on the point of moving.

It should be stressed that the above laws are experimental and are accepted as the basis for the mathematical treatment of friction. Modern theory suggests that the force of friction is in fact due to the non-rigidity of bodies. When one body rests on another there is always an area of contact which is much smaller than the apparent area and also depends on the normal pressure between the bodies. Friction is considered to be due to the fusion of materials (of which the bodies are composed) over the area of contact. Therefore friction would be proportional to the area of contact and therefore proportional to the normal pressure as assumed in the above laws.

Equilibrium Analysis of Simple Systems with Sliding Friction

When the inertia forces are neglected in comparison to the externally applied load one may go for static force analysis. If the body is under equilibrium condition then this equilibrium is known as static equilibrium and this condition is applicable in many machines where the movement is relatively slow. These include clamps, latches, support linkages and many hand operated tools such as pliers and cutters. In case of lifting cranes also the bucket load and the static weight loads may be quite high relative to any dynamic loads due to accelerating masses and hence one may go for static force analysis.

When the inertia effect due to the mass of the components is also considered it is called dynamic force analysis.

Applied and Constraint forces:

- When two or more bodies are connected together to form a group or system the pair of action and reaction forces between any two of the connecting bodies is called constrained forces.
- These forces constrain the connected bodies to behave in a specific manner defined by the nature of the connection.
- Forces acting on this system of bodies from outside the system are called applied forces. Electric, Magnetic and gravitational forces are example of forces that may be applied without actual physical contact. But most of the forces we are concerned in mechanical equipment occur through direct physical or mechanical contact.

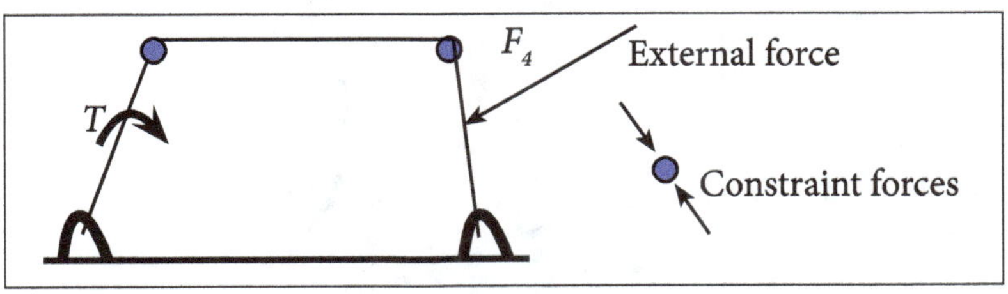

It Constraint forces of action and reaction at a mechanical contact occur in pairs and thus have no net force effect on the system of bodies being considered. When a part of the body is considered in isolation the effect of such force is considered by using the free body diagram. Characteristics of a force are its magnitude, its direction and its point of application.

Two equal and opposite forces along two parallel but non-collinear straight lines in a body cannot be combined to constitute a single force and they constitute a couple. The arm of the couple is the perpendicular distance between their lines of action and the plane of the couple is the plane containing the two lines of action.

The moment of the couple M is a vector directed normal to the plane of the couple and the sense of M is in accordance to the right-hand rule for rotation.

The moment of couple:

$$M = R_{BA} \times F$$

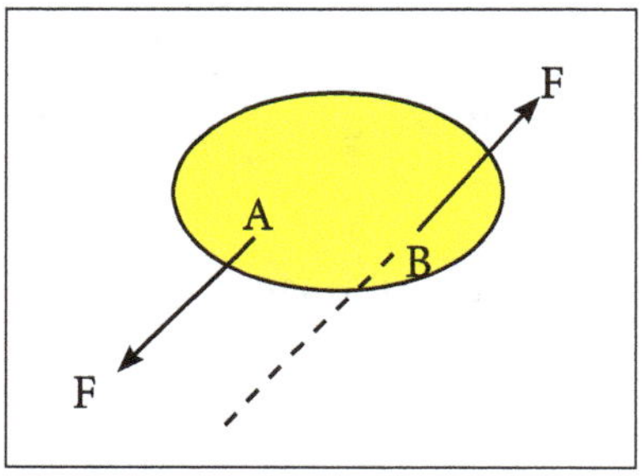

The value of M is independent of the choice of the reference point about which the moments are taken because the vector is the same for all positions of the origin. R_{BA} As the moment vector M is independent of any particular origin or line of application hence it is a free vector.

Wedge Friction

A wedge is in general a triangular object which is placed between two objects to either hold them in place or is used to move one relative to the other. For example the following shows a wedge under a block that is supported by the wall.

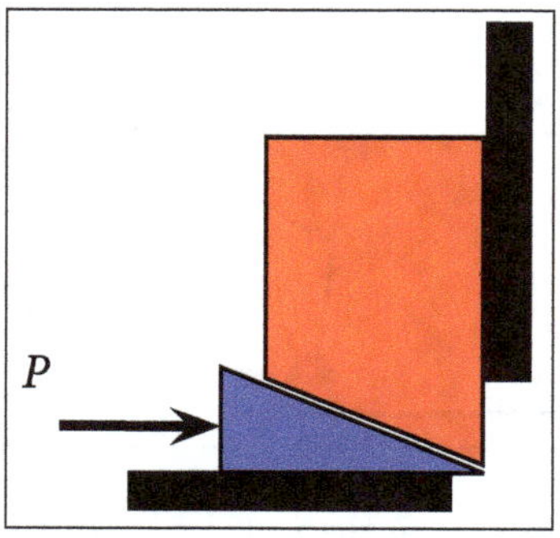

If the force P is large enough to push the wedge forward then the block will rise and the following is an appropriate free-body diagram. Note that for the wedge to move one needs to have slip on all three surfaces. The direction of the friction force on each surface will oppose the slipping.

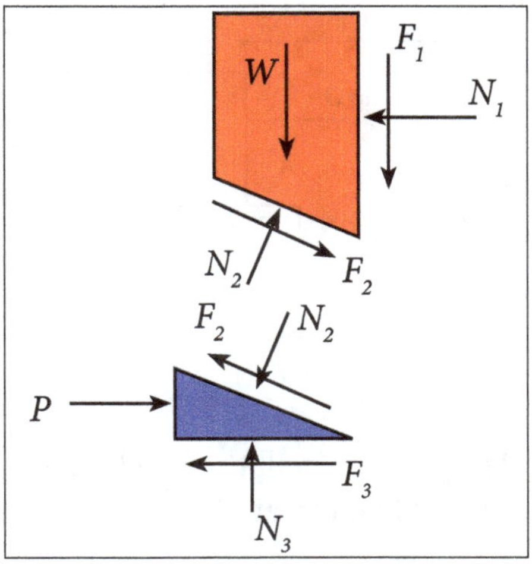

Since before the wedge can move each surface must overcome the resistance to slipping one can assume that,

$$F_1 = \mu N_1$$
$$F_2 = \mu N_2$$
$$F_3 = \mu N_3$$

These equations and the equations of equilibrium are combined to solve the problem. If the force P is not large enough to hold the top block from coming down, then the wedge will be pushed to the left and the appropriate free-body diagram is the following. Note that the only change is the direction of the frictional forces. A similar analysis to the above yield the solution to the problem.

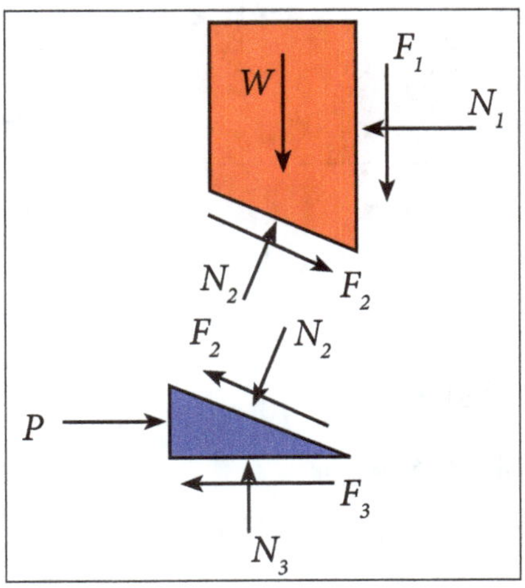

Problems

1. A Two blocks A and B of mass 50 kg and 100 kg respectively are connected by a string C which passes through a friction less pulley connected with the fixed wall by another string D as shown in figure.

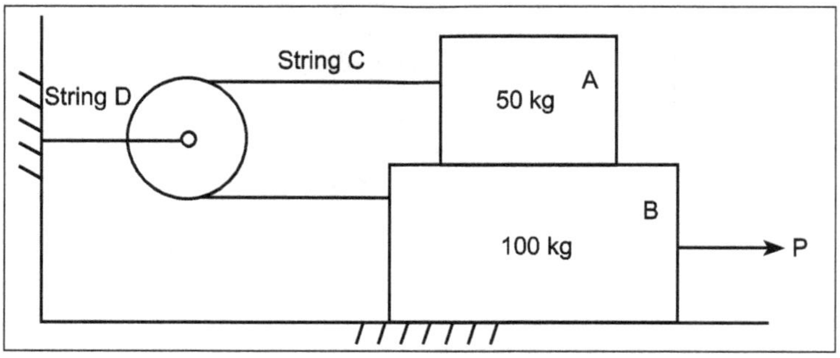

Let us determine the force P required to pull the block B. And also calculate the tension in the string D. Take coefficient of friction at all contact surfaces as 0.3.

Solution

Given data:

$m_A = 50kg$

$m_B = 100\ kg.$

Solution:

Free body diagram at Block A and B:

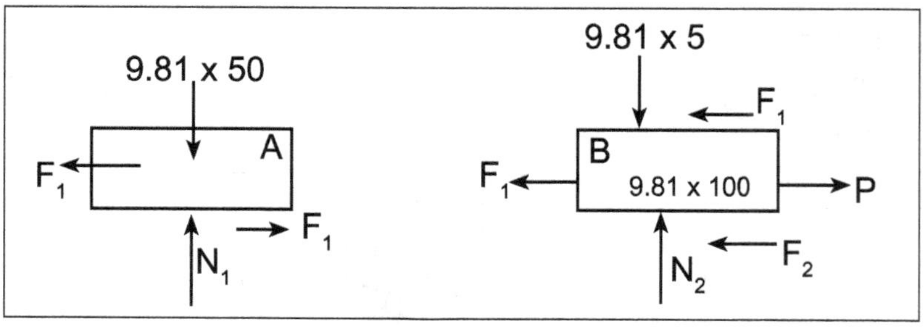

Consider Block A:

$$N_1 = 50 \times 9.81 = 490\ N$$

Where F_i – Friction force on block A.

Pulling force acting on the block A is J_1:

$$T_1 = F_1 = 0.3 \times 490.5 = 147.15N.$$

Consider Block B ($\sum V = 0$)

$$N_2 = (50+100)\times 9.81 = 1471.5$$

$$N_2 = 1471.5\ N$$

WKT:

$$F_2 = v_2 \times N_2$$

$$P = T_1 + F_1 + F_2$$

$$P = 147.15 + 147.15 + (0.3\times 1471.5)$$

$$P = 735.75\ N.$$

Tension in String D′ =T_2:

$$T_2 = T_1 + T_1$$

$$= 147 + 147 = 294.3.$$

Result:

Force in the String D = 294 N.

2. The static coefficient of friction μ_s between the block shown in Figure having a mass of 75 kg and the surface. Let us calculate the magnitude and direction of the friction force if the force P applied is inclined at 45° to the horizontal and μ_s = 0.30.

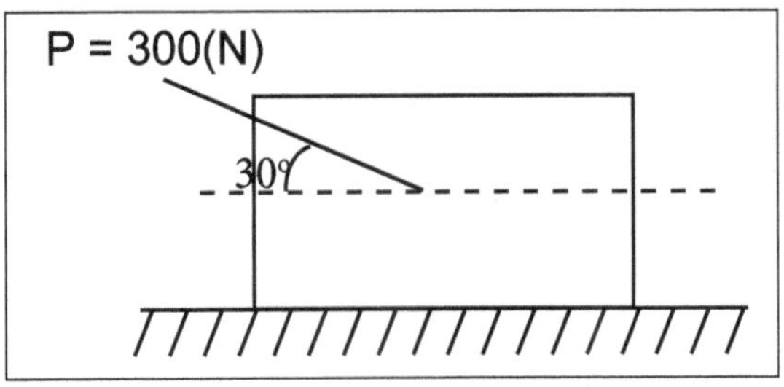

Solution

Given data:

Case-1: Mass of the block = 75kg.

Weight of the block = 75 x 9.81 = 735.75 N

Force applied, P = 300 N

Inclination of the force with horizontal = 30°

Co-efficient of friction, μ_s = ?

Case-2: Inclination of the force, = 45°

Co-efficient to friction, μ_s= 0.30

Magnitude of frictional force = ?

Direction of frictional force = ?

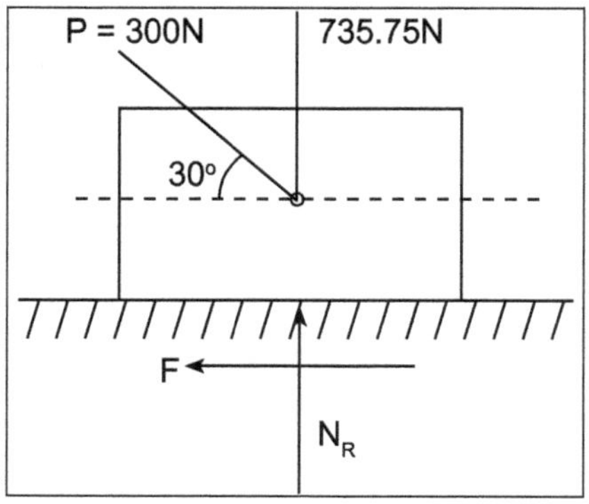

Case-1:

$$N_R - 300 \sin 30° - 735.75 = 0$$
$$N_R = 885.75 \text{ N}$$
$$F = \mu N_R = 885.75\ \mu$$
$$F = 885.75\ \mu$$
$$300 \cos 30° - F = 0$$
$$300 \cos 30° - 885.75\mu = 0$$
$$\mu = 0.293.$$

Case-2:

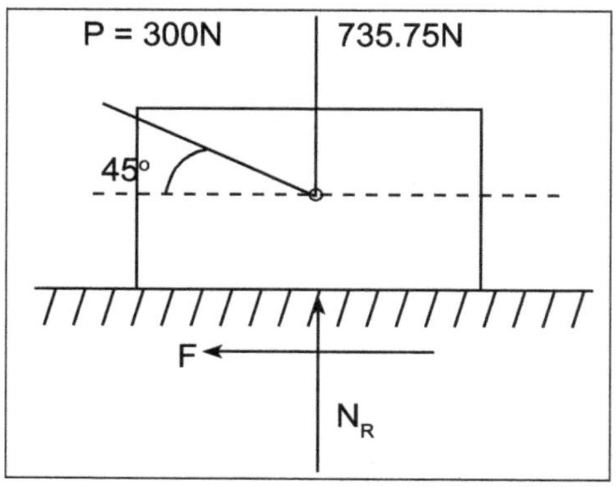

$$N_R - 300 \sin 45° - 735.75 = 0$$
$$N_R = 947.88 \text{ N}$$

$$F = \mu N_R = 0.3(947.88)$$
$$F = 284.36 \text{ N.}$$

Answer:

$$\mu = 0.293$$

$$F = 284.36\text{N.}$$

3. The block A in figure supports a load W and is to be raised by forcing the wedge B under it. If the angle of friction is 10° at all surfaces in contact let us determine the maximum wedge angle α that will give the wedge a mechanical advantage; i.e., make P less than the weight W of the block.

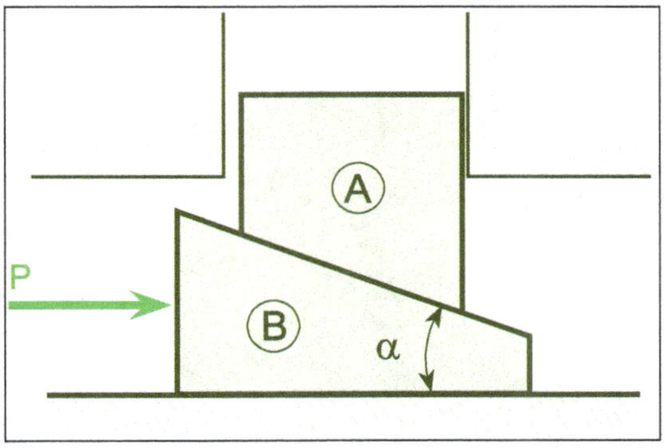

$$\beta = 180° - 100° - (10° + \alpha)$$
$$\beta = 70° - \alpha$$
$$\frac{R_2}{\sin 100°} = \frac{W}{\sin\beta}$$

$$\frac{R_2}{\sin 100°} = \frac{W}{\sin(70° - \alpha)}$$
$$R_2 = \frac{W \sin 100°}{\sin(70° - \alpha)}$$

$$\theta = 180° - 80° - (80° - \alpha)$$
$$\theta = 20° + \alpha$$
$$\frac{R_2}{\sin 80°} = \frac{P}{\sin\theta}$$
$$R_2 = \frac{P \sin 80°}{\sin\theta}$$
$$\frac{W \sin 100°}{\sin(70° - \alpha)} = \frac{W \sin 80°}{\sin(20° - \alpha)}$$

$$\frac{\sin 100^\circ}{\sin(70^\circ - \alpha)} = \frac{\sin 80^\circ}{\sin(20^\circ - \alpha)}$$

$$\sin 100^\circ \sin(20^\circ - \alpha) = \sin 80^\circ \sin(70^\circ - \alpha)$$

$$\sin 100^\circ (\sin 20^\circ \cos \alpha + \cos 20^\circ \sin \alpha) = \sin 80^\circ (\sin 70^\circ \cos \alpha - \cos 70^\circ \sin \alpha)$$

$$\sin 100^\circ \sin 20^\circ \cos \alpha + \sin 100^\circ \cos 20^\circ \sin \alpha = \sin 80^\circ \sin 70^\circ \cos \alpha - \sin 80^\circ \cos 70^\circ \sin \alpha$$

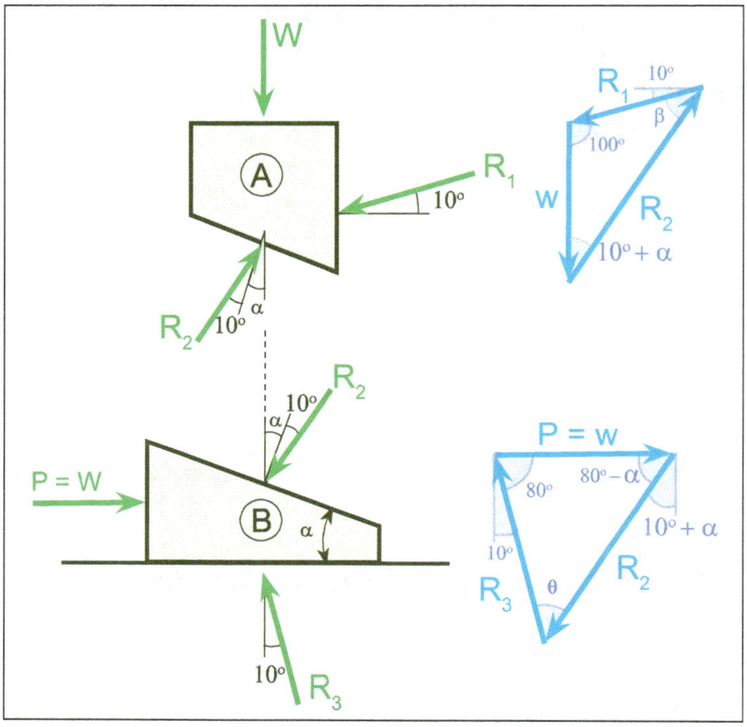

$$\sin 100^\circ \cos 20^\circ \sin \alpha + \sin 80^\circ \cos 70^\circ \sin \alpha = \sin 80^\circ \sin 70^\circ \cos \alpha - \sin 100^\circ \sin 20^\circ \cos \alpha$$

$$(\sin 100^\circ \cos 20^\circ + \sin 80^\circ \cos 70^\circ) \sin \alpha = (\sin 80^\circ \sin 70^\circ - \sin 100^\circ \sin 20^\circ) \cos \alpha$$

$$\frac{\sin \alpha}{\cos \alpha} = \frac{\sin 80^\circ \sin 70^\circ - \sin 100^\circ \sin 20^\circ}{\sin 100^\circ \cos 20^\circ + \sin 80^\circ \cos 70^\circ}$$

$$\tan \alpha = \frac{\sin 80^\circ \sin 70^\circ - \sin 100^\circ \sin 20^\circ}{\sin 100^\circ \cos 20^\circ + \sin 80^\circ \cos 70^\circ}$$

$$\tan \alpha = 0.4663076582$$

Answer: $\alpha = 25^\circ$

5

Centre of Gravity and Moment of Inertia

5.1 Centroid and Centre of Gravity

Centroid

The center of gravity of a body is defined as the point through which the entire weight of the body acts. When this is referred to weightless lamina's or plane area it is called the centroid of the area.

Difference between Centroid and Centre of Mass

The centroid is purely a geometrical. It is the centre of gravity for objects of uniform density. However an object's density can be non-uniform, which will move the centre of gravity away from the centre. Imagine a 2D shape made from cardboard. Here the point in the middle where we can balance it on the end of a pin is the centroid. But if we then replace part of the shape with steel, keeping the same shape, the centroid is the same but the centre of gravity has shifted.

Centroids of Area

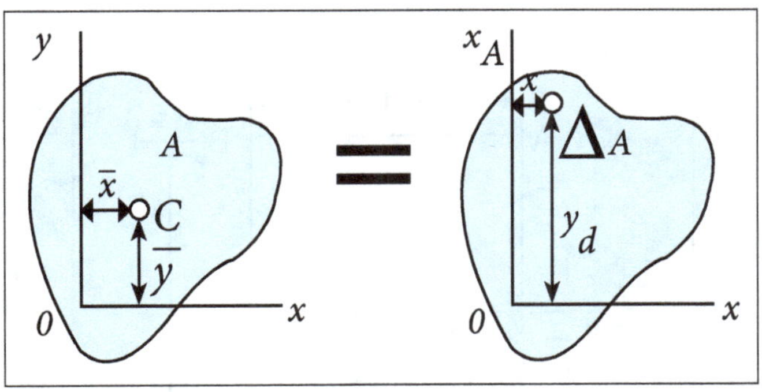

$$\bar{x}A = \sum_{i=1}^{n} \bar{x}_i A_i$$

$$= \int_A \bar{x}_{el} dA$$

$$\bar{y}A = \sum_{i=1}^{n} \bar{y}_i A_i$$
$$= \int_A \bar{y}_{el} dA$$

Centroids of Line

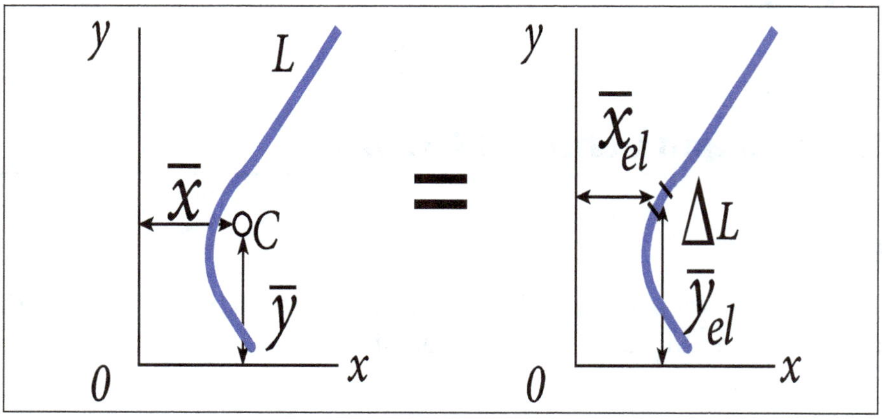

$$\bar{x}L = \int_L \bar{x}_{el} dL$$
$$\bar{y}L = \int_L \bar{y}_{el} dL$$

Centroid of a Rectangular Section by Integration

The figure.1 shows a rectangular section ABCD having width = b and depth = d. Consider a rectangular elementary strip of thickness 'dy' at a distance y from the axis OX.

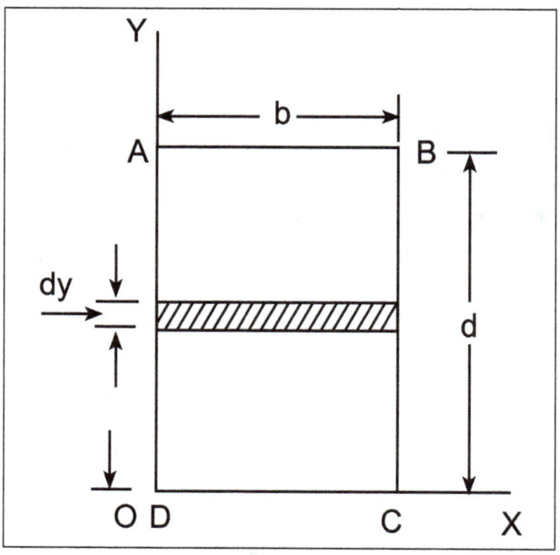

Figure: 1

Let dA = Area of strip=b.by

Moment of the area dA about axis OX

$$= dA \times y = (b.dy) \times y \qquad (\because dA = b.dy)$$
$$= by \times dy$$

The moment of the whole area about axis OX (or first moment of the whole area about axis OX) will be obtained by integrating the above equation between the limits o to d.

$$\therefore = \int_o^d by \times dy = b\int_o^d y\,dy$$

($\because$ b is constant and can be taken outside the integral sign):

$$= b\left[\frac{y^2}{2}\right]_o^d = \frac{bd^2}{2} \qquad ...(i)$$

Let A=Total area of rectangular section:

$$= \int_o^d dA = \int_o^d b\,dy = b[y]_o^d \qquad [\because dA = b.dy]$$

$\bar{y}$ = distance of the centroid of the rectangular section from axis OX.

$\therefore$ Moment of total area of rectangular section about axis OX:

$$= A \times \bar{y} = (b \times d) \times \bar{y} \qquad ...(ii)$$

Equating the equations (i) and (ii), we get:

$$(b \times d) \times \bar{y} = \frac{bd^2}{2}$$

$$\therefore \bar{y} = \frac{bd^2}{2} \times \frac{1}{bd} = \frac{d}{2}$$

Similarly, the distance of the centroid of the rectangular section from the axis OY is given by:

$$\bar{x} = \frac{b}{2}$$

Refer to Figure (a):

Area of strip, $dA = d.\ dx$

Moment of this area dA about axis OY,

$$= dA \times x$$

$\therefore$ Moment of the whole area about axis OY:

$$= \int_0^b dA \times x = \int_0^b (d.dx)x$$

$[\because dA = d.dx]$

$$= d\int_0^b x\,dx = d\left[\frac{x^2}{2}\right]_0^b = \frac{d \times b^2}{2} \quad ...(iii)$$

Let $\bar{x}$ =Distance of centroid of whole area from axis OY.

$$A = \text{Total area} = b \times d$$

$\therefore$ Moment of total area A about axis OY:

$$= A \times \bar{x}$$

Equating equation (iii) and (iv), we get:

$$A \times \bar{x} = \frac{b \times d^2}{2}$$

$$\text{or } \bar{x} = \frac{b \times d^2}{2} \times \frac{1}{A}$$

$$= \frac{b \times d^2}{2} \times \frac{1}{b \times d} = \frac{b}{2}$$

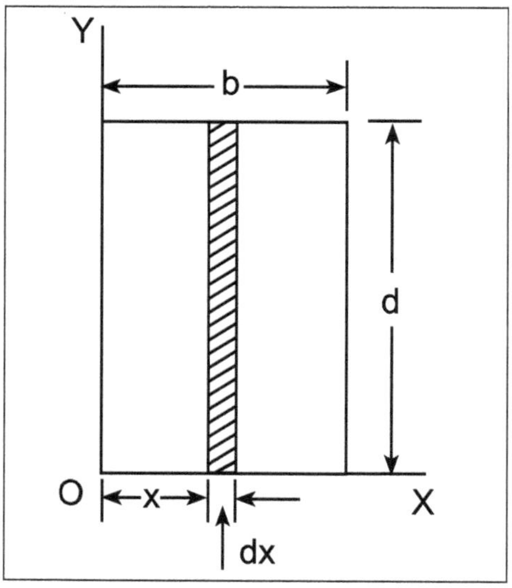

Figure: 1(a).

Centroid of Circular Section by Integration

The figure.2 shows a circular section of radius R with o as centre. The equation of the circle is given by:

$$x^2 + y^2 + R^2$$

Consider a rectangular elementary strip of thickness dy at a distance of y from the axis OX.

The area of strip, dA=2x.dy

Moment of this area dA about x –axis

$$= dA.y$$
$$= (2x.dy).y \ (\because dA = 2x.dy)$$
$$= 2xy.dy \qquad ...(i)$$

But the equation of the circle is,

$$x^2 + y^2 + R^2$$
$$\text{or } x = \sqrt{R^2} - y^2$$

Substituting the above value of x in equation (i), we get

Moment of area dA about x-axis

$$= 2\sqrt{R^2} - y^2.y.dy$$

Moment of total area A about x-axis will be obtained by integrating the above

Equation form –R to R

∴ Moment of area A about x –axis

$$= \int_{-R}^{+R} 2\sqrt{R^2 - y^2}\,(-2y).dy$$
$$= \left[\frac{\left(R^2 - y^2\right)^{3/2}}{3/2}\right]_{-R}^{R}$$
$$= \frac{2}{3}\left[\left(R^2 - R^2\right)^{3/2} - \left\{\left(R^2 - R^2\right)^{3/2}\right\}\right]$$
$$= \frac{-2}{3}[0 - 0] = 0 \qquad ...(ii)$$

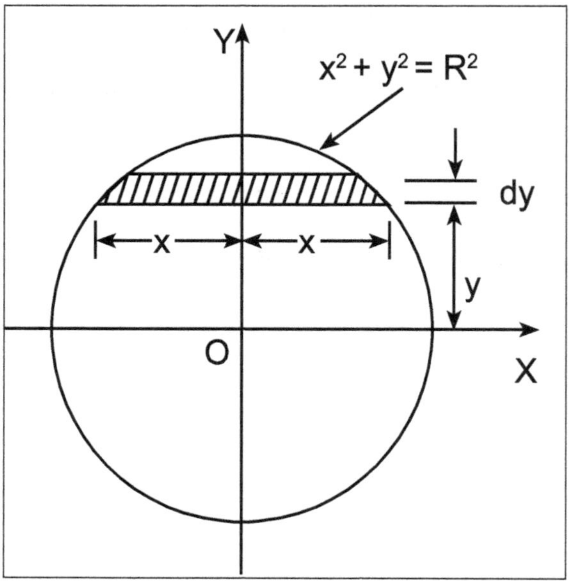

Figure: 2

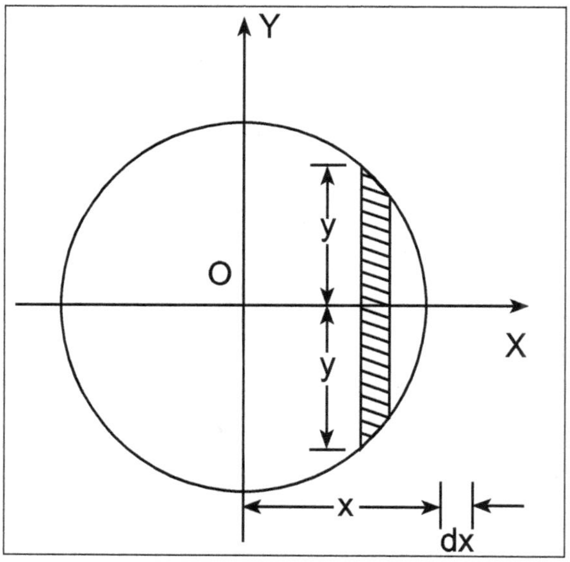

Figure:2a

Also the moment of total area A about $x-\text{axis} = A \times \bar{y}$ ...(iii)

Where $\bar{y}$ =Distance of centroid of total area A from $x-\text{axis}$.

Equating the two values given by equation (ii) and (iii), we get: $A \times \bar{y} = 0$ or $\bar{y} = 0$

This means that centroid of the circle is at the centre of the circle.

Centroid of a Triangular Section by Integration

The figure.3 shows a triangular section AOB of base width = b and height = h.

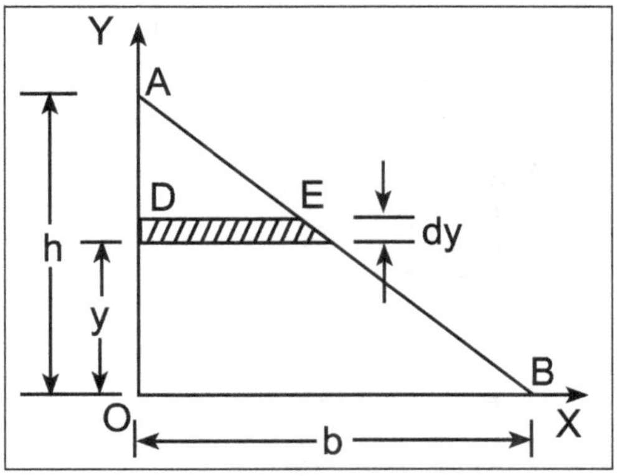

Figure: 3

Consider a small strip of thickness dy at a distance y from the axis OX.

Area of strip, $dA = \text{Length} \times DE$ ...(i)

The distance DE in terms of y, b and h is obtained from two similar triangles ADE and AOB as:

$$\frac{DE}{OB}=\frac{AD}{AO}$$

Where, $OB = b, AO = h$ and $AD = (h-y)$

$$\therefore \frac{DE}{b}=\frac{(h-y)}{h}$$

Or $$DE=\frac{b(h-y)}{h}$$

Substituting this value of DE in equation (i), we get:

Area of strip, $dA=\frac{b(h-y)}{h}.dy$

Moment of this area dA about axis OX:

$$dA.y=\frac{b(h-y)}{h}.dy.y$$

$$=\frac{b}{h}(h-y).y.dy$$

The moment of the total area A of the triangular section is obtained by integrating the above equation between the limits o to h.

$$=\frac{b}{h}\left[h^2\frac{h^2}{2}\right]=\frac{b}{h}\cdot\frac{h^2}{2}\frac{bh}{2}$$

Then moment of total area A about axis OK $\Lambda\times\bar{y}=\left(\frac{b\times h}{2}\right)\times\bar{y}$ ...(iii)

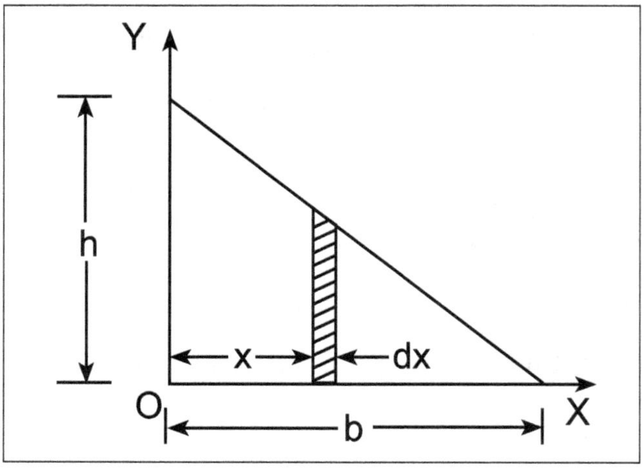

Figure: 3(a)

Equating the two values given by equations (i) and (iii), we get:

$$=\left(\frac{b\times h}{2}\right)\times\bar{y}$$

$$=\frac{b}{h}\cdot\frac{h^3}{6}$$

$$\text{or } \bar{y}=\left(\frac{b}{h}\cdot\frac{h^3}{6}\right)\times\frac{1}{\left(\frac{b\times h}{2}\right)}=\frac{h}{3}$$

Similarly, taking a strip of thickness dx parallel to y-axis at a distance of x from the y-axis [Refer to Figure.3 (a)}, it can be proved that:

$$\bar{x}=\frac{b}{3}.$$

Shapes	Images	$\bar{x}$	$\bar{y}$	Area
Triangular Area	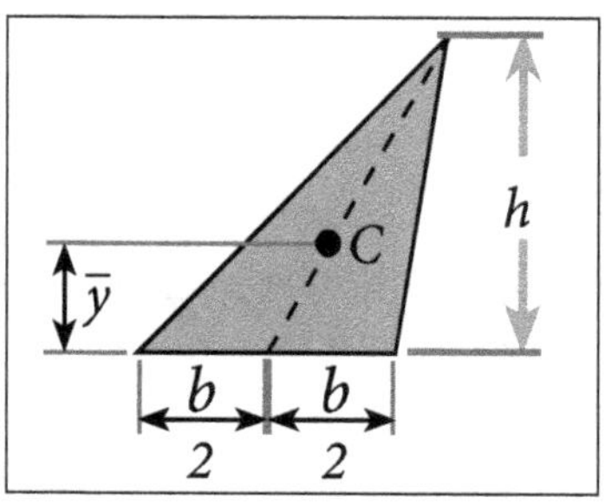		$\frac{h}{3}$	$\frac{bh}{2}$

<table>
<tr><td>Quarter–circular area</td><td rowspan="2">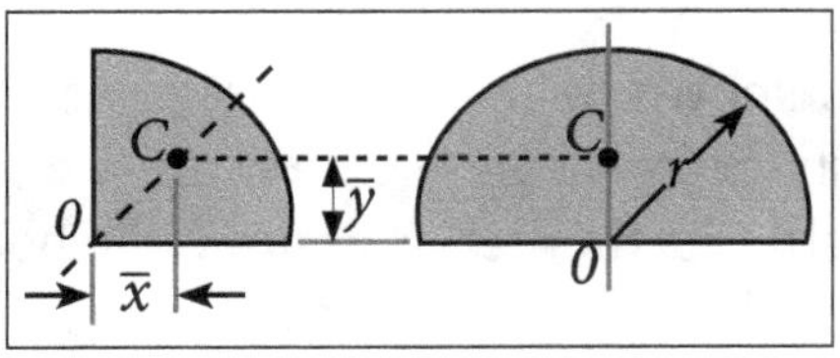</td><td>$\frac{4r}{3\pi}$</td><td>$\frac{4r}{3\pi}$</td><td>$\frac{\pi r^2}{3\pi}$</td></tr>
<tr><td>Semicircular area</td><td>0</td><td>$\frac{4r}{3\pi}$</td><td>$\frac{\pi r^2}{2}$</td></tr>
<tr><td>Semi parabolic area</td><td rowspan="2">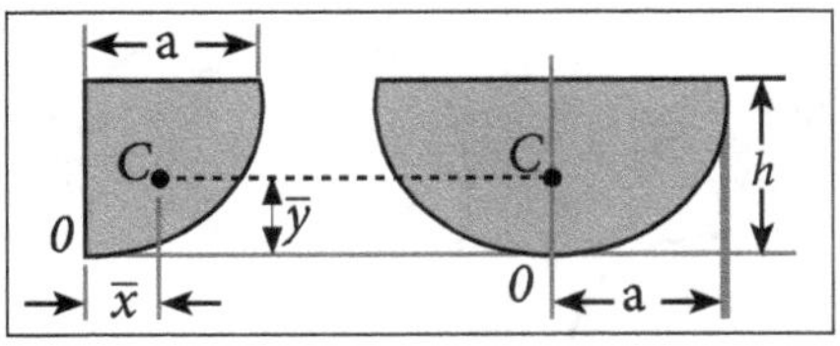</td><td>$\frac{3\alpha}{8}$</td><td>$\frac{3h}{5}$</td><td>$\frac{2\alpha h}{3}$</td></tr>
<tr><td>Parabolic area</td><td>0</td><td>$\frac{3h}{5}$</td><td>$\frac{4\alpha h}{3}$</td></tr>
<tr><td>Parabolic spandrel</td><td>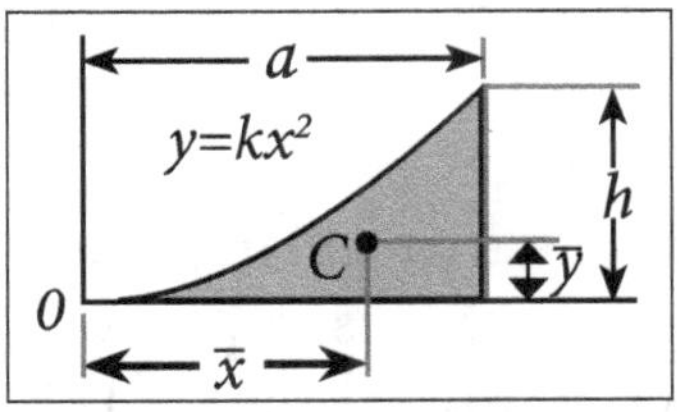</td><td>$\frac{3\alpha}{4}$</td><td>$\frac{3h}{10}$</td><td>$\frac{\alpha h}{3}$</td></tr>
<tr><td>Circular Sector</td><td>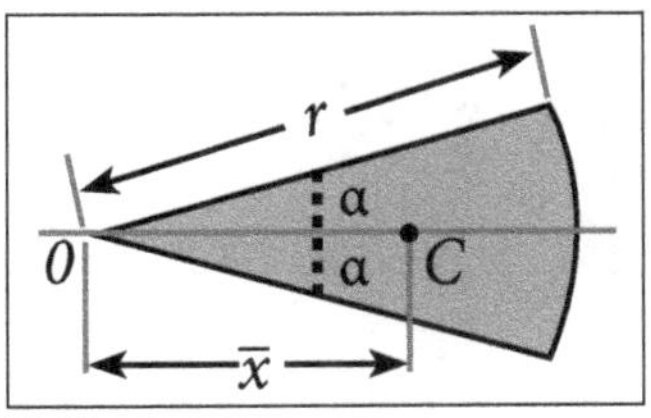</td><td>$\frac{2r \sin\alpha}{3\alpha}$</td><td>0</td><td>$\alpha r^2$</td></tr>
<tr><td>Quarter-circular arc</td><td rowspan="2">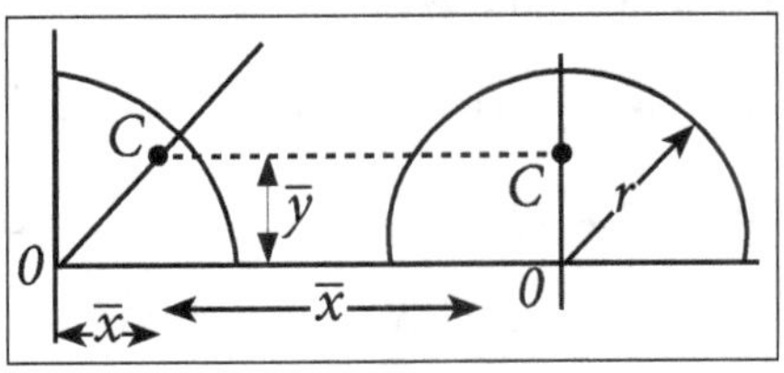</td><td>$\frac{2r}{\pi}$</td><td>$\frac{2r}{\pi}$</td><td>$\frac{\pi r}{2}$</td></tr>
<tr><td>Semicircular arc</td><td>0</td><td>$\frac{2r}{\pi}$</td><td>πr</td></tr>
<tr><td>Arc of circle</td><td>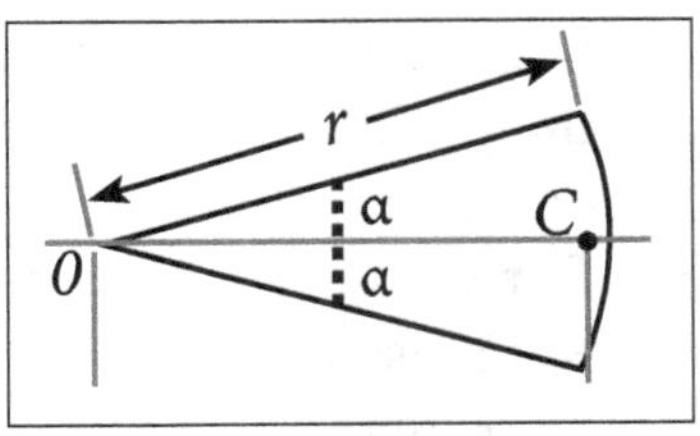</td><td>$\frac{r \sin\alpha}{\alpha}$</td><td>0</td><td>$2\alpha r$</td></tr>
</table>

Centre of Gravity

Consider the following lamina shown below. Let us assume that it has been exposed to gravitational field. Obviously every single element must experience a gravitational force towards the centre of earth. Further let us assume the body has practical dimensions, then we shall easily conclude that all elementary forces will be unidirectional and parallel.

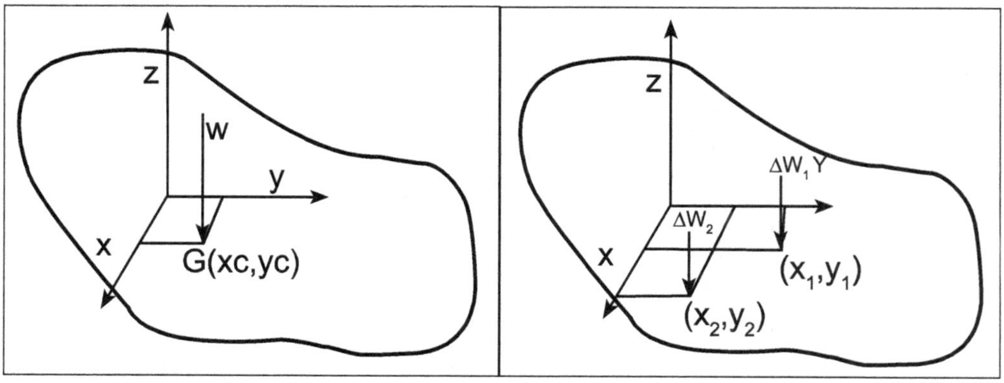

Let us consider G to be the centroid of the irregular lamina. We can easily represent the net force passing through the single point G as shown in the above first figure. We can also divide the entire region into say n small elements. Let us say the coordinates to be as shown in figure such as:

$$(x_1, y_1), (x_2, y_2), (x_3, y_3) \ldots\ldots\ldots (xn, yn)$$

Let ΔW_1, ΔW_2, ΔW_3,......., ΔW_n be the elementary forces acting on the elementary elements. Obviously,

$W = \Delta W_1 + \Delta W_2 + \Delta W_3 + \ldots\ldots\ldots\ldots + \Delta W_n$When n tends to infinity ΔW it becomes infinitesimally small and it shall be replaced as dW.

The Centre of gravity is said to be,xc= $\int x\ dw / \int dw$,yc= $yc = \int y\ dw$ /and $zc = \int z\ dw / \int dw$ in case of a three dimensional body)where dw(or ΔW) the elemental force and x,y are the coordinate of the small element. And we have seen that $\int dw = W$

For certain type of surfaces of bodies there lies a probability that the centre of gravity may lie outside the body. Secondly it is said that the centre of gravity represents the entire lamina, thus we can replace the entire body by a single point with the force acting on it when required. There is a major difference between the centre of mass and the centre of gravity of a body. For the centre of gravity we shall integrate with respect to dW whereas for the centre of mass we integrate with respect to dm. Force is a vector quantity and mass is a scalar quantity. For general practical size objects both of them turns out to be the same for both of them where they are proportional and the force is unidirected (dW = dm*g). But when we consider large size objects such as a continent, the results would be different because here the vector nature of dW comes into play.

Centre of Gravity of a Composite Body

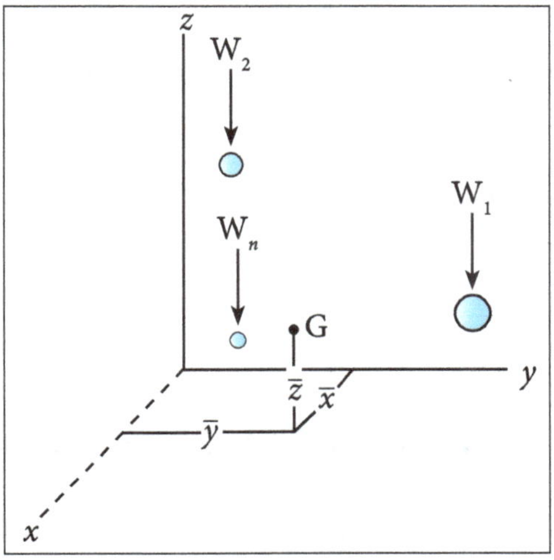

Consider a composite body which consists of a series of particles (or bodies) as shown in the figure.

The net or the resultant weight is given as WR = ΣW.

By summing the moments about the y-axis, we get:

$$\bar{x} W_R = \tilde{x}_1 W_1 + \tilde{x}_2 W_2 + \ldots\ldots\ldots\ldots + \tilde{x}_n W_n$$

Where $\tilde{x}_1$ represents x coordinate of W_1, and same as follows.Similarly, we can sum the moments about the x- and z-axes to find the coordinates of G.

By replacing the W with a M in these equations, the coordinates of the center of mass can be found.

Polar Moment of Inertia

The Polar moment of inertia is a measure of a circular beam's ability to resist torsion. It is required to calculate the turn of a beam subjected to a torque. It is analogous to the space of the moment of inertia, which characterizes a beam's ability to resist bending and is required to calculate displacement. The long of the polar moment of inertia, the minimum of the beam will twist, when subjected to a given torque.

The polar moment of inertia must not be confused with the moment of inertia, which characterizes an object's angular acceleration due to a torque.

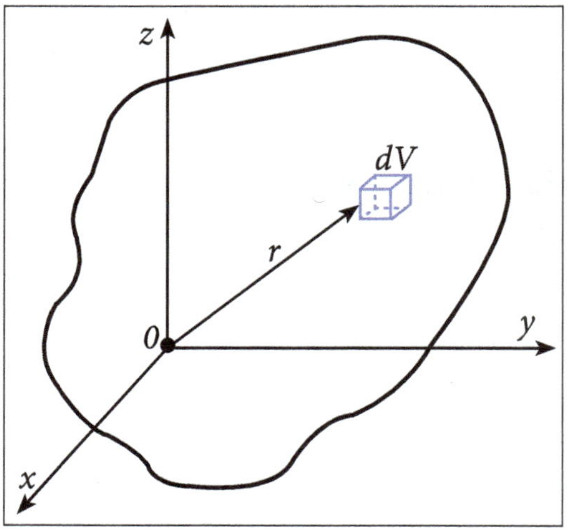

5.2 Moment of Inertia of Area and Mass, Radius of Gyration

Moment of Inertia

According to the first law of motion given by Newton says that a body remains in its current motion state until any external force is acted upon it. The moment of inertia of the linear motion can be measured by the mass of the system and angular counterpart. The moment of inertia is related to the amount of mass distribution throughout the body and the mass. It is not necessary to be that it must possess same moment of inertia by two bodies of the same mass. Therefore, the inertia shows that how tough a body to move. If an object possess more inertia then it is very to move it or stop it. For example, the inertia of the heavy truck is more than a person. It also shows how hard it is to spin an object with respect to certain axis that depends on the mass and the location of mass. Thus, the spinning difficulty around an axis is said to be the moment of inertia.

Consider the distributed forces whose magnitudes are proportional to its elemental areas on which they act and also they vary linearly with the distance of from a given axis.

Leonhard Euler introduced the concept of moment of inertia. The resistance that is shown by the object to change its rotation is termed as moment of inertia. The symbols I and J are used for denoting the moment of inertia. The moment of inertia describes the angular acceleration that is produced by an applied torque. It depends on the shape, mass distribution of the body and also on the orientation of the rotational axis. The moment of inertia plays the same role for rotation same as the mass does for a translational motion, where it describes the resistance of a body to a change of its state of motion.

The difficulty that is faced due to change the angular motion of any object about an axis is shown or given or measured by calculating moment of inertia about that particular axis.

The term "moment of inertia" is used very frequently in mechanical design calculations. There are two different types of moment of inertia:

- Mass moment of inertia.
- Area moment of inertia.

Moment of Inertia Equation

Moment of inertia is given by the following equation, $I = \sum^{N}_{0} m_i r_i^2$ Here ri is the distance to the axis on which rotation is done to any other point and not distance of the point to the origin. Similarly, if we replace the summation by integral then we can find the moment of inertia of a continuous solid body rotating about a particular known axis:

$$I = \sum \int \text{volume}\, p(r) d(r) dV(r)$$

where, r = Radius vector P (r) = Mass density d (r) = Distance from a given point r to the known axis of rotation.

There are two theorems for moment of inertia:

1. Parallel Axis Theorem:

$$I_d = I_{com} + Md^2.$$

Where, d = distance between a axis that is parallel and an axis passing through the center of mass and the it is the perpendicular distance M = the total mass of the object or body.

I_d = moment of inertia about the parallel axis considered I_{com} = the moment of Inertia about the axis that is basically passing through Center of Mass (com).

2. Perpendicular Axis Theorem:

$$I_k = I_i + I_j$$

The three mutually perpendicular axes are i, k and j.

Area Moment of Inertia

The second moment of the area around a given axis is called the area moment of inertia.

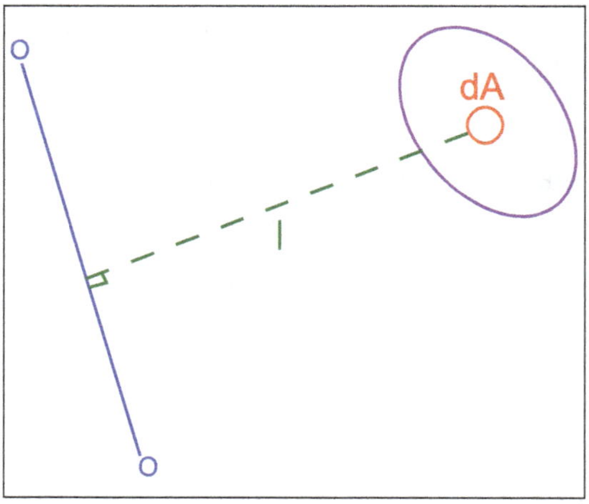

Now, for the above figure we have the axis o given and hence we can calculate the moment of area by summing together l2 dA for all the given elements of area dA in the yellow region. Hence Area moment of inertia, $I = \int A l^2 dA$ ...(a)

For a rectangular region the area moment of inertia, $I_x = \int A\, y^2 dA Iy = \int A\, x^2 dA$.

Mass Moment of Inertia

The mass moment of inertia is the measurement of the distribution of the mass of an object or body relative to a given axis. It is represented by the symbol I. The formula for a single particle of mass m is, $I_o = MR^2$.

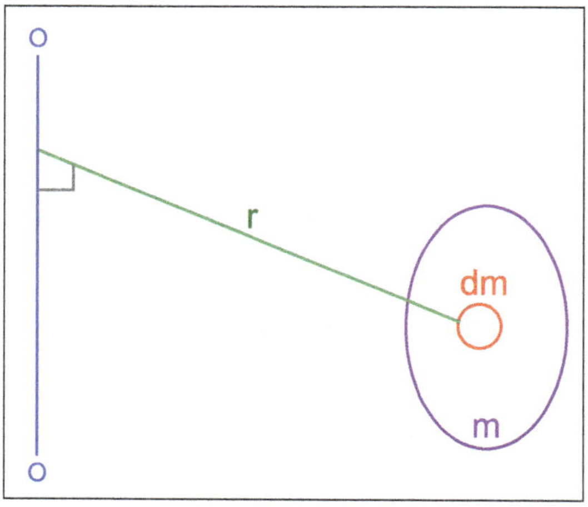

Units: kg/m^2

Mass moment for rigid bodies,

Here the body is thought of as sum of particles each having mass dm. The formula is:

$$I = \int \sum m r^2 dm$$

Similarly:

$$I = \int \Sigma r^2 \ p \, pV$$

Where, p = density

If calculated over area then:

$$I = \int \Sigma r^2 p \, dA$$

Moments of Inertia of Composite Figures

A similar theorem can be used with the polar moment of inertia. The polar moment of inertia J_o of an area about O and the polar moment of inertia JC of the area about it's given by:

Types of section	Moment of inertia	Y max	Section modules (z)
Rectangle or paralleogram 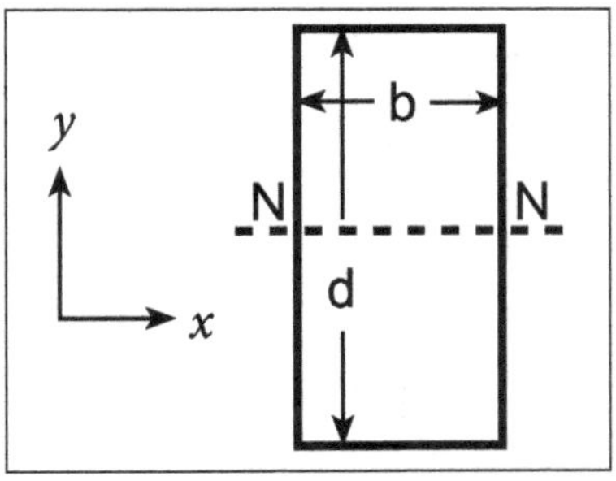	$I_{xx} = \frac{bd^3}{12}$ $I_{yy} = \frac{db^3}{12}$	$\frac{d}{2}$ $\frac{b}{2}$	$Z_{XX} = \frac{bd^2}{6}$ $Z_{yy} = \frac{db^2}{6}$
Hollow rectangular section 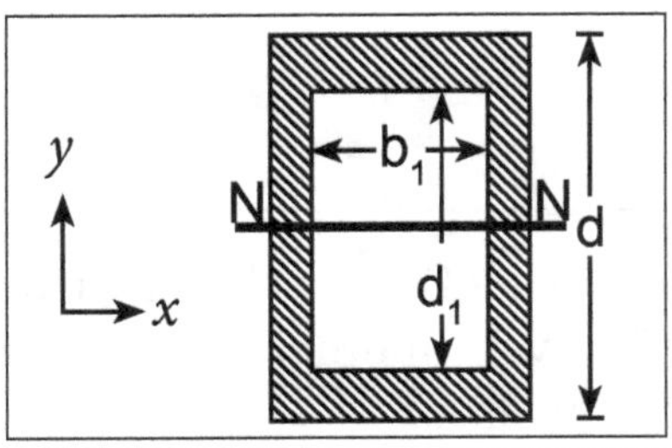	$I_{xx} = \frac{bd^3}{12} - \frac{b_1 d_1^3}{12}$ $I_{yy} = \frac{db^3}{12} - \frac{d_1 b_1^3}{12}$	$\frac{d}{2}$ $\frac{b}{2}$	$Z_{xx} = \frac{1}{6d}\left(bd^3 - b_1 d_1^3\right)$ $Z_{yy} = \frac{1}{6b}\left(db^3 - d_1 b_1^3\right)$
Cercular section 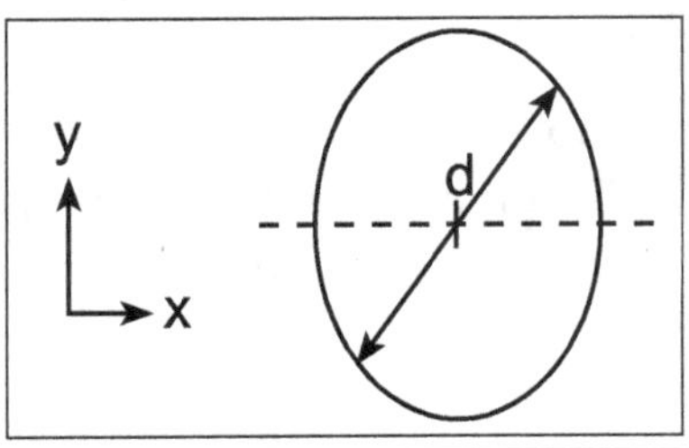	$I_{xx} = \frac{p}{64} d^4$ $I_{yy} = \frac{p}{64} d^4$	$\frac{d}{2}$ $\frac{d}{2}$	$Z_{xx} = \frac{p}{32} d^3$ $Z_{yy} = \frac{p}{32} d^3$

Hollow curcular section 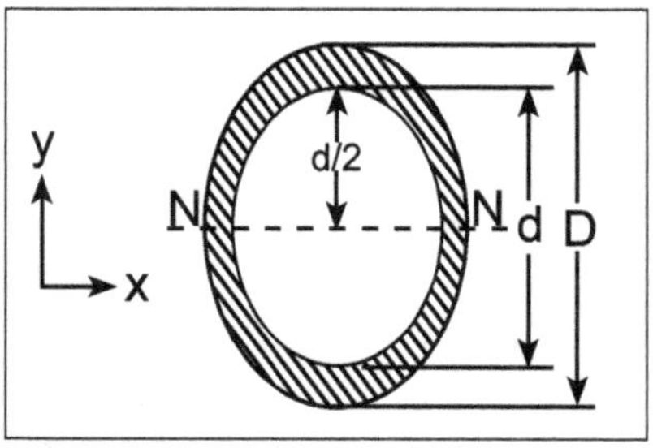	$I_{xx} = I_{yy} = I$ $I_{yy} = \frac{p}{64}\left(D^4 - d^4\right)$	$\frac{D}{2}$	$Z_{xx} = Z_{yy} = Z$ $Z = \frac{p}{32D}\left(D^4 - d^4\right)$
I-section 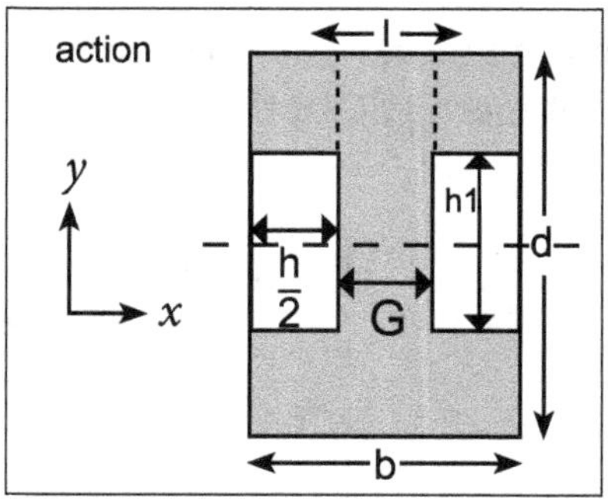	$I_{xx} = \frac{bd^3}{12} - \frac{b_1 d^3_1}{12}$ $I_{yy} = \frac{db^3}{12} - \frac{d_1 b^3_1}{12}$ or $I_{xx} = \frac{1}{12}\left(bd^3 - (b-t)d^3_1\right)$	$\frac{d}{2}$ $\frac{b}{2}$	$Z_{xx} = \frac{1}{6d}\left(bd^3 - b_1 d^3_1\right)$ $Z_{yy} = \frac{1}{6b}\left(db^3 - d_1 b^3_1\right)$
Triangle 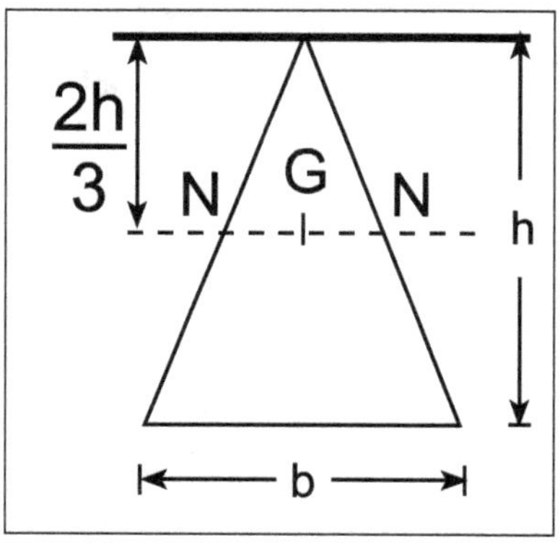	$I_G = \frac{bh^3}{36}$	$\frac{2}{3}h$	$Z_G = \frac{bh^2}{24}$

Its centroid are related to the distance d between two points C and O by the relationship. The parallel-axis theorem is being used very effectively in order to compute the moment of inertia of a composite area with respect to a given axis.

The Mass Moment of Inertia for a Particle

The one measure of the distribution of the mass of an object relative to a given axis. is said to be the mass moment of inertia for a particle. The mass moment of inertia is denoted by I and is given for a single particle of mass m as given shown below:

$$I_o = r^2 m$$

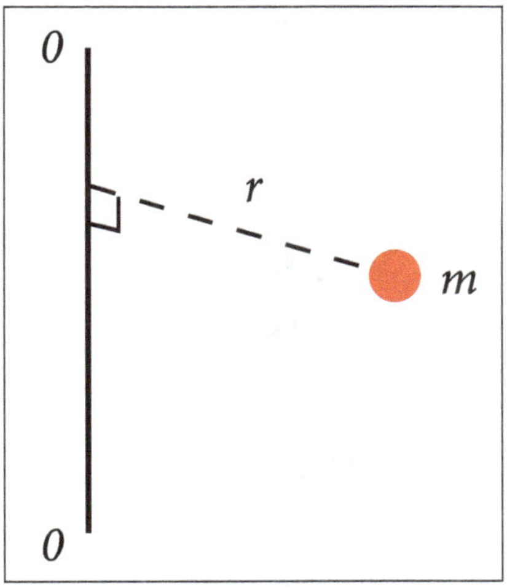

Where O-O is the axis around which one is evaluating the mass moment of inertia and r is the perpendicular distance between the axis O-O and the mass . As seen from the above equation, the mass moment of inertia has the units of mass times the length squared.

The mass moment of inertia should not be confused with the area moment of inertia that has units of length to the power four. It naturally appear in the equations of motion and provide information on how difficult. The moment of inertia rotates the particle around given axis.

The application in rigid body dynamics measure of distribution of mass of a rigid body w.r.t. the axis (constant property for that axis).

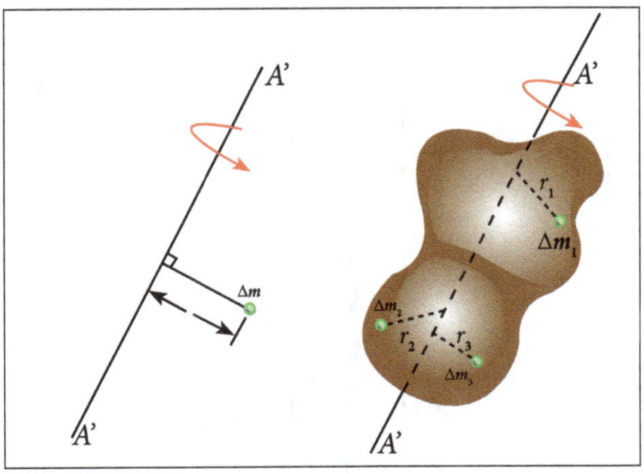

$$I = \int r^2 dm$$

r = Perpendicular distance of the mass element dm from the axis O-O.

r2Δm: Measure of the inertia of the system.

About Individual Coordinate Axes

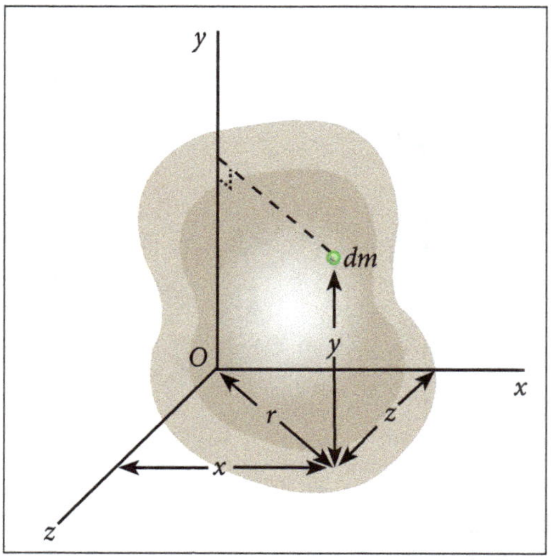

$$I_x = \int \left(y^2 + z^2\right) dm$$
$$I_y = \int \left(z^2 + x^2\right) dm$$
$$I_z = \int \left(x^2 + y^2\right) dm$$

Parallel Axis Theorem

$$x = x' + \bar{x} \qquad y = y' + \bar{y} \qquad z = z' + \bar{z}$$

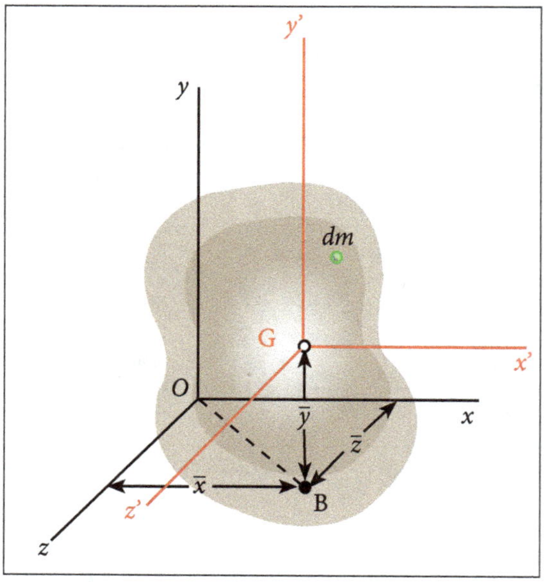

$$I_x = \bar{I}_{x'} + m\left(\bar{y}^2 + \bar{z}^2\right)$$

$$I_y = \overline{I}_{y'} + m\left(\overline{z}^2 + \overline{x}^2\right)$$

$$I_z = \overline{I}_{z'} + m\left(\overline{x}^2 + \overline{y}^2\right)$$

Moments of Inertia of Thin Plates

A thin plate of uniform thickness t and homogeneous material of density r, the mass moment of inertia with respect to the axis AA' which contained in the plate is:

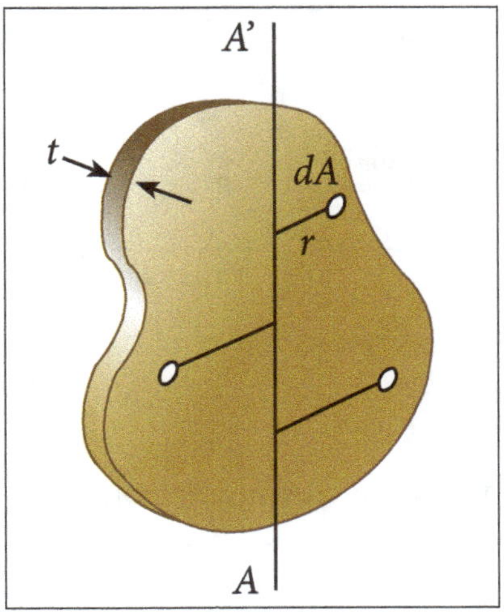

$$I_{AA'} = \int r^2 dm = pt\int r^2 dA$$
$$= pt\,I_{AA',area}$$

Similarly, for the perpendicular axis BB' which is also contained in the plate:

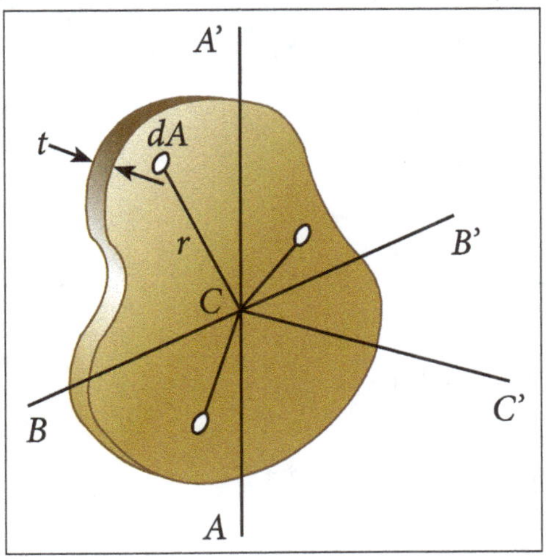

$$I_{BB'} = pt\,I_{BB',area}$$

For axis CC' which is perpendicular to the plate:

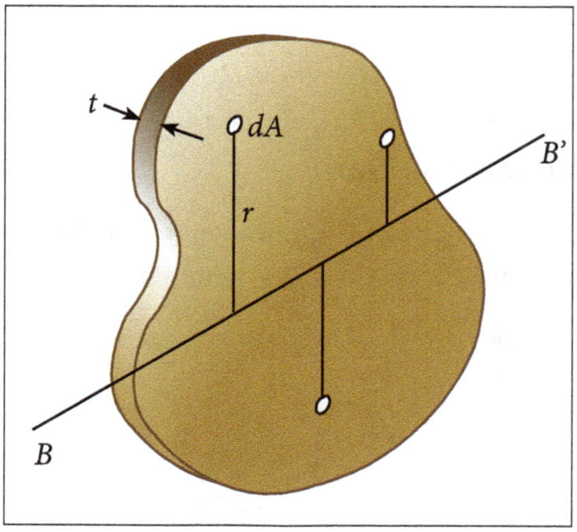

$$I_{CC'} = \rho t\, J_{C,area} = \rho t\left(I_{AA',area} + I_{BB',area}\right)$$
$$= I_{AA'} + I_{BB'}$$

For the principal centroidal axes on a rectangular plate:

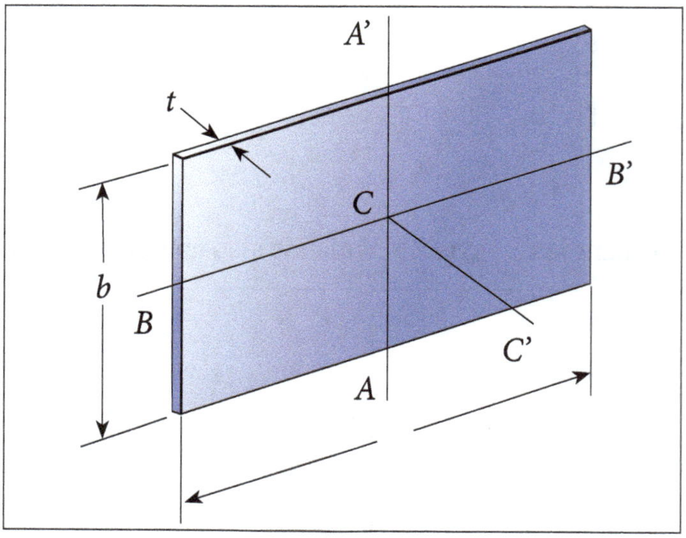

$$I_{AA'} = \rho t\, I_{AA',area} = \rho t\left(\frac{1}{12}a^3 b\right) = \frac{1}{12}ma^2$$

$$I_{BB'} = \rho t\, I_{BB',area} = \rho t\left(\frac{1}{12}ab^3\right) = \frac{1}{12}mb^2$$

$$I_{CC'} = I_{AA',mass} + I_{BB',mass} = \frac{1}{12}m\left(a^2 + b^2\right)$$

For the centroidal axes on a circular plate:

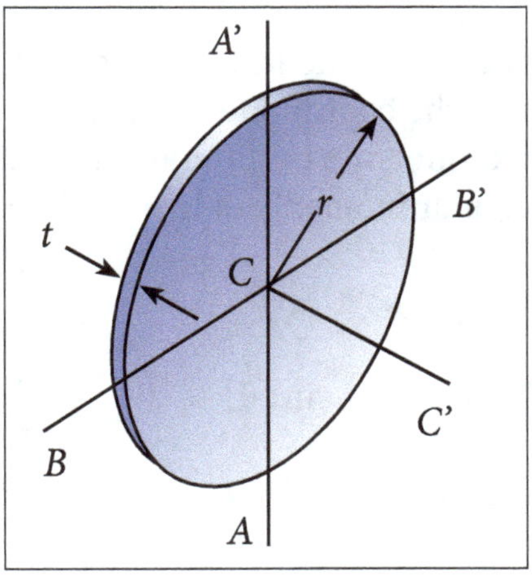

$$I_{AA'} = I_{BB'} = pt\,I_{AA',area} = pt\left(\frac{1}{4}\pi r^4\right)$$
$$= \frac{1}{4}mr^2$$

Radius of Gyration

$$k_x = \sqrt{\frac{I_x}{A}}$$
$$k_y = \sqrt{\frac{I_y}{A}}$$

The radius of gyration of an area with respect to a particular axis is the square root of the quotient of the moment of inertia divided by the area. It is said to be the distance at which the entire area must be assumed to be concentrated in order that the product of the area and the square of this distance shall equal the moment of inertia of the actual area about the given axis. Therefore the radius of gyration can also be described as the way in which the total cross-sectional area is distributed around its centroidal axis. If the area is more distributed further from the axis it will have greater resistance to buckling. The most efficient column section in order to resist buckling is a circular pipe because it has its area distributed as far away as possible from the centroid. Rearranging we have:

$$I_x = k_x^2 A$$
$$I_y = k_y^2 A$$

The radius of gyration is the distance k away from the axis that all the area can be concentrated to result in the same moment of inertia.

5.3 Introduction to Product of Inertia and Principle Axes

Thus the sum of paired rectangular moments of inertia about a pole is always a constant regardless of the orientation of the coordinate axes. However the rectangular moment is a function of angle θ, a maximum or a minimum value can be obtained by differentiating either one of the rectangular moments of inertia. Imply:

$$\because \frac{dI_u}{d\theta} = 0 \text{ or } \frac{dI_v}{d\theta} 0$$

$$\rightarrow \begin{cases} I_u = \left(I_x \cos^2\theta + I_y \sin^2\theta - I_{xy} \sin 2\theta\right) \\ I_v = \left(I_x \sin^2\theta + I_y \cos^2\theta + I_{xy} \sin 2\theta\right) \end{cases}$$

$$\begin{cases} \therefore \frac{dI_u}{d\theta} = \frac{d}{d\theta}\left(I_x \cos^2\theta + I_y \sin^2\theta - I_{xy} \sin 2\theta\right) \\ \Rightarrow \frac{dI_u}{d\theta} = -2I_x \sin\theta \cos\theta + 2I_y \sin\theta \cos\theta - 2I_{xy} \cos 2\theta \\ \therefore \frac{dI_v}{d\theta} = \frac{d}{d\theta}\left(I_x \sin^2\theta + I_y \cos^2\theta + I_{xy} \sin 2\theta\right) \\ \Rightarrow \frac{dI_v}{d\theta} = -2I_x \sin\theta \cos\theta - 2I_y \sin\theta \cos\theta + 2I_{xy} \cos 2\theta \end{cases}$$

$$\because \frac{dI_u}{d\theta} = 0 \text{ or } \frac{dI_v}{d\theta} = 0 \rightarrow \theta = \theta_m$$

$$\Rightarrow \begin{cases} \frac{dI_u}{d\theta} = -2I_x \sin\theta_m \cos\theta_m + 2I_y \sin\theta_m \cos\theta_m - 2I_{xy} \cos 2\theta_m = 0 \\ \frac{dI_v}{d\theta} = -2I_x \sin\theta_m \cos\theta_m - 2I_y \sin\theta_m \cos\theta_m + 2I_{xy} \cos 2\theta_m = 0 \end{cases}$$

$$\Rightarrow \left(I_y - I_x\right) \sin 2\theta_m = 2I_{xy} \cos 2\theta_m$$

$$\Rightarrow \tan 2\theta_m = \frac{2I_{xy}}{I_y - I_x} = -\frac{2I_{xy}}{I_x - I_y}$$

$$\Rightarrow \tan 2\theta_m = \frac{2I_{xy}}{I_x - I_y}$$

$$\Rightarrow -\theta_m = \frac{1}{2} \tan^{-1} \frac{2I_{xy}}{I_x - I_y}$$

Through the differentiation angle θ of the rectangular coordinate axes at which the rectangular moments of inertia is either maximum or minimum can be determined.

Those paired rectangular coordinate axes is termed as principal axes and the paired rectangular moments of inertia is known principal moments of inertia.

The origin O of the rectangular coordinate axes can be located either inside or outside the area if the origin O is chosen to coincide with the centroid the two principal axes of the area about the axes through its centroid C are termed as the principal centroidal axes of the area.

Since a tangent function is said to be a periodic function with period π the equation defines two values 2θm of 180° apart and the two values of θ_m are 90° apart that is confirmed with the maximum and minimum values of rectangular moments of inertia about the rectangular coordinate axes that is the principal moments of inertia about O. Imply,

$$\Rightarrow \tan 2\theta_m = \frac{2I_{xy}}{I_y - I_x}$$

$$\Rightarrow \tan(2\theta_m + n\pi) = -\frac{2I_{xy}}{I_y - I_x}; n = 0,1$$

$$\Rightarrow \tan\left(2\left(\theta_m + n\frac{\pi}{2}\right)\right) = -\frac{2I_{xy}}{I_y - I_x}; n = 0,1$$

5.4 Supports, Shear Force, Bending Moment Diagram and Loads

Type of Supports:

Roller Support

The roller supports are free to rotate and translate along the surface upon which the roller rests. The surface can be horizontal, vertical or slopped at any angle. Roller supports are commonly located at one end of long bridges in the form of bearing pads.

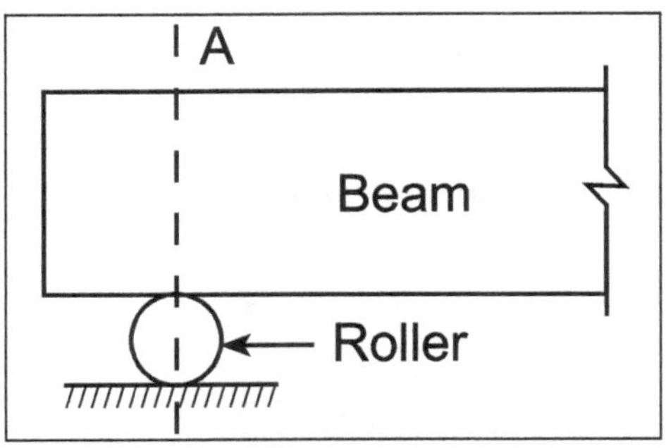

Roller support

Hinge Support

The hinge support is capable of the resisting forces acting in any direction of the plane. Here, this support does not provide any resistance to rotation. The horizontal and vertical component of reaction shall be determined using equation of equilibrium.

Hinge support may also be used in the three hinged arched bridges at the banks supports while at the center internal hinge is introduced. It is also used in doors in order to produce only one rotation in a door. The hinge support reduces sensitivity to earthquake.

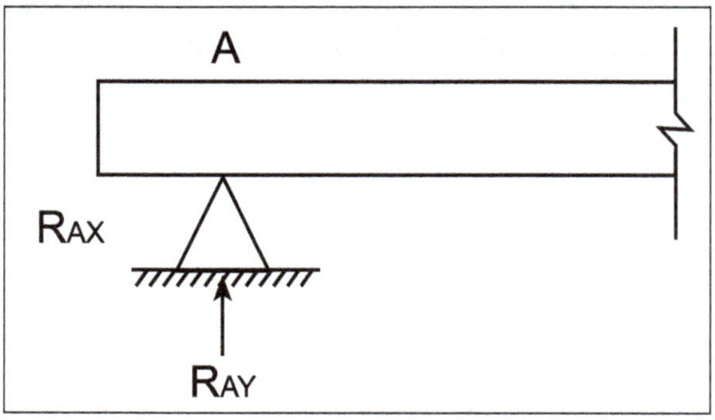

Hinge support.

Fixed Support

The fixed support can resist vertical and horizontal forces as well as moment since they restrain both rotation and translation. They are also called as rigid support. For stability of a structure there should be one fixed support.

A flagpole at concrete base is a common example of fixed support In RCC structures the steel reinforcement of a beam. Similarly, all the riveted and welded joints in steel structure are the examples of fixed supports Riveted connection are not very much common now a days due to the introduction of bolted joints.

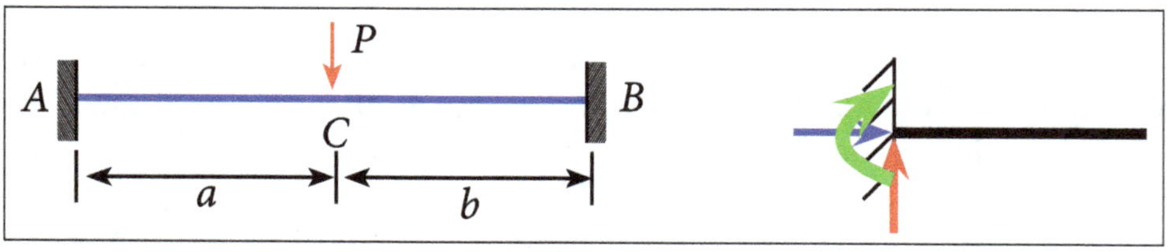

Fixed support.

Shear Force and Bending Moment Diagram for Cantilever and Simply Supported Beam

Beam

Beam is a structural member which is supported along the length and subjected to external loads acting transversely (i.e) perpendicular to the center line of the beam.

Transverse Loading on Beam

If a load is acting on the beam which is perpendicular to the central line of it then it is called transverse loading.

Types of beams:

Cantilever Beam

A beam in which one end is free and the other end is fixed is called cantilever beam.

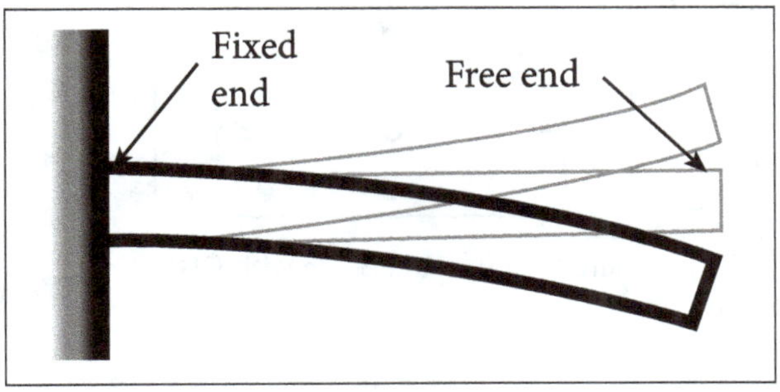

Cantilever Beam.

Simply Supported Beam

A beam supported or resting free on the support at its both ends.

BM Formula for UDL

Max BM = wL2/8.

Maximum Bending Moment

The bending moment is max. Writing SF equation at that point and equating to zero we can find out the distances 'x' from one end then find maximum bending moment at that point by taking all moment on right or left hand side of beam.

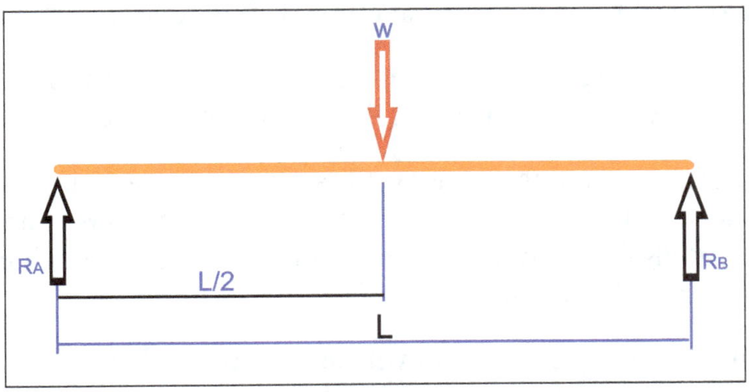

Simply Supported Beam with point load.

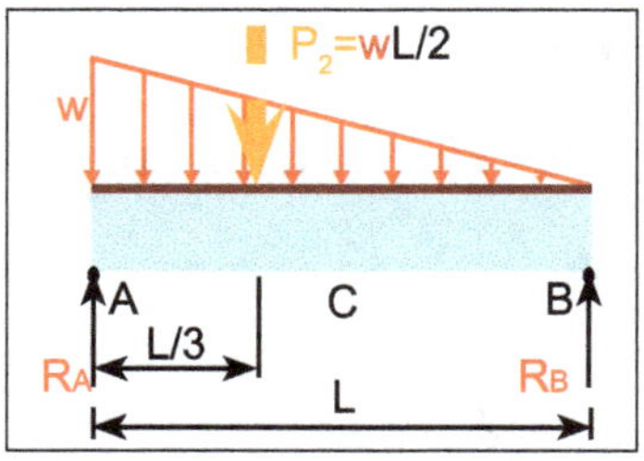

Simply Supported Beam with UVL.

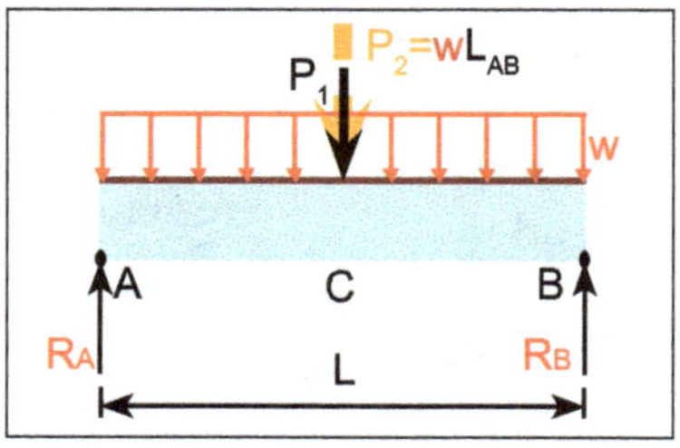

Simply Supported Beam with UDL.

Types of Loads

- Concentrated load: A concentrated load is said to be the load which is acting at a point is called point load.
- Uniformly Varying Load: A load which is spread over a beam in such a manner that rate of loading varies from point to point but at uniformly varying is known as uniformly varying load, briefly written as Uniformly Varying Load u.v.l.
- Uniformly distributed load: If a load which is spread over a beam in such a manner that rate of loading 'w' is uniform throughout the length then it is called as uniformly distributed load (udl).
- Point of contra-flexure: Point at which BM changes to zero is point of contra-flexure. It occurs in overhanging beam.
- Positive or sagging BM: BM is said to positive if moment on left side of beam is clockwise or right side of the beam is counter clockwise.
- Negative or hogging BM: BM is said to negative if moment on left side of beam is counterclockwise or right side of the beam is clockwise.
- Shear force and bending moment: SF at any cross section is defined as algebraic sum of all the forces acting either side of beam. BM at any cross section is defined as algebraic sum of the moments of all the forces which are placed at either side from that point.
- Maximum bending moment: BM will be maximum when shear force change its sign.

Problems

1. Here we draw SFD and BMD for point load.

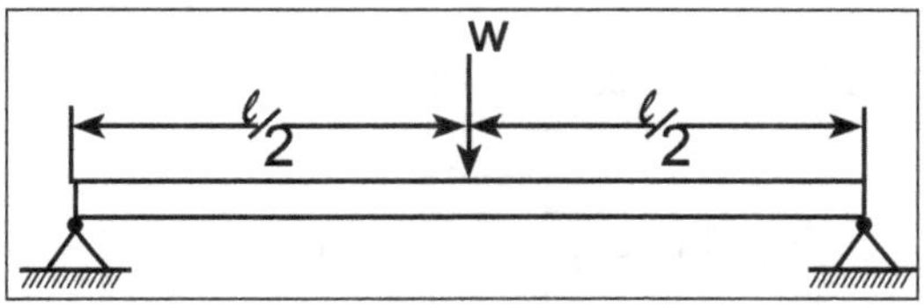

By symmetry the reactions at the two supports might be W/2 and W/2. Then consider any section X-X from the left end then, the beam is under the action of following forces.

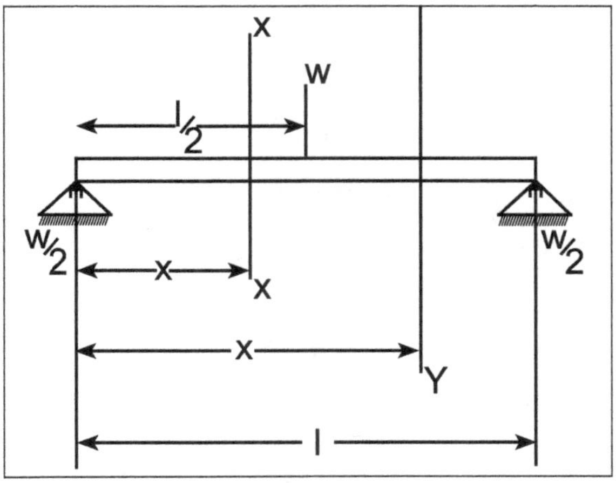

Therefore the shear force at any X-section would be = W/2 [Which is constant upto x < l/2].

If we consider another section say Y-Y which is beyond l/2 then:

$$S.F_{Y.Y} = \frac{W}{2} - w = \frac{-W}{2}$$

For all values greater = l/2

Hence the S.F diagram can be plotted as:

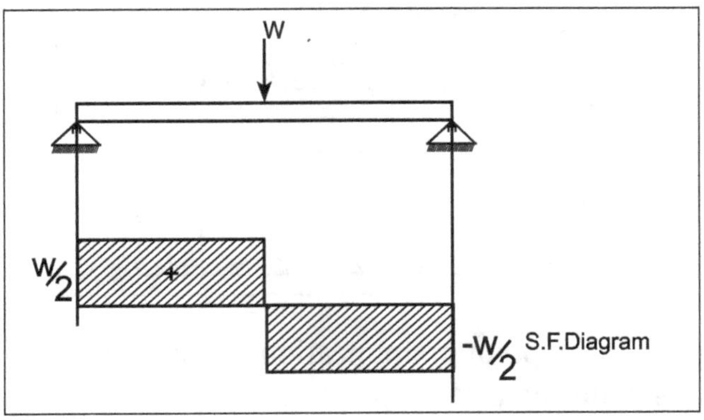

For B.M Diagram

If we just take the moments to the left of the cross-section by:

$$B.M_{X.X} = \frac{w}{2} x \text{ for x lies between o and ½}$$

$$B.M_{at\,x=\frac{1}{2}} = \frac{w}{2}\frac{1}{2} \text{ i.e B.M at x = o}$$

$$= \frac{Wl}{4}$$

$$B.M_{Y.Y} = \frac{W}{2}x - W\left(x - \frac{1}{2}\right)$$

Again

$$= \frac{W}{2}x - Wx + \frac{Wl}{2}$$

$$= -\frac{W}{2}x + \frac{Wl}{2}$$

$$B.M_{at\,x-l} = -\frac{Wl}{2} + \frac{Wl}{2}$$

$$= 0$$

Which when plotted will give a straight relation, i.e.

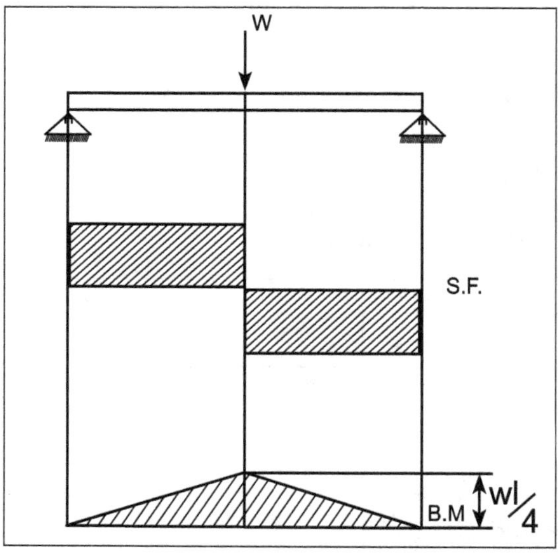

It can be observed that at the point of application of load there is an abrupt change in the shear force, at this point the B.M is maximum.

2. A simply supported beam overhanging on one side is subjected to a U.D.L. of 1 kN/m. Let us sketch the shear force and the bending moment diagrams and find the position of point of contra flexure.

Solution:

Consider a section (X – X') at a distance x from end C of the beam.

To draw the SFD and BMD we need R_A and R_B:

By taking the moment of all the forces about point A.

$$R_B \times 3 - w/2 \times (4)^2 = 0.\ R_B = 1 \times (4)^2 / 2 \times 3 = 8/3 \text{ kN}.$$

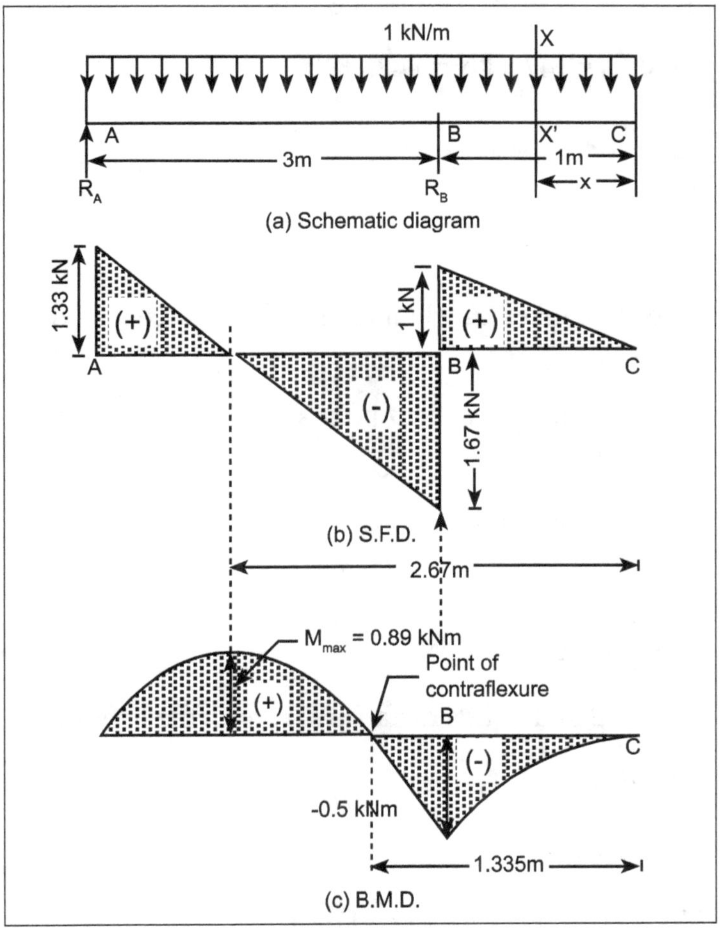

For static equilibrium:

$$R_A + R_B - 1 \times 4 = 0.$$

$$R_A = 4 - 8/3 = 4/3 \text{ kN}.$$

To draw SFD we need S.F. at all salient points:

By taking a section between C and B, SF at a distance x from end C. we have,

$$Fx = +w.x \text{ kN}$$

At x = 0; F_C = 0.

x = 1 m; F_B just right = 1 × 1 = + 1 kN.

Now, by taking a section between B and A, at a distance x from end C, the SF is:

$$Fx = x - 8/3$$

When, $x = 1$ m; $FB = 1 - 8/3 = -5/3$ kN $= -1.67$ kN.

At $x = 4$ m; $FA = 4 - 8/3 = +4/3$ kN $= +1.33$ kN.

The S.F. becomes zero:

$$Fx = x - 8/3 = 0.$$

$$\Rightarrow x = 2.67 \text{ m}.$$

(The sign is taken positive taken as when the resultant force is in downward direction on the RHS of the section).

To draw BMD we need B.M. at all salient points. Taking section between C and B, BM at a distance x from end C, we have:

$$Mx = -\ wx^2/2 = -1.x^2/2 \text{ kNm}.$$

When x = 0; M_C = 0.

At $x = 1$ m; $MB = -1 \times (1)2/2 = -0.5$ kNm.

Taking section between B and A, at a distance x from C, the bending moment is:

$$Mx = -x^2/2 + 8/3\ (x - 1)$$

At $x = 1$ m; $MB = -0.5$ kNm.

$$x = 4 \text{ m};\ MA = -(4)2/2\ - 8/3\ (4 - 1) = 0.$$

The maximum bending moment occurs at a point where:

$$dMx/dx = 0.$$

$$\Rightarrow d/dx\left[-x^2/2\ + 8/3x - 8/3\right] = 0.$$

$\Rightarrow -1/2 \times 2x + 8/3 = 0.$

$\Rightarrow x = 8/3$ m from end C

$\Rightarrow M_{max} = -1/2\ (8/3)2 + 8/3(8/3 - 1) = 0.89$ kNm.

The point of contra flexure occurs at a point, where:

$Mx = 0.$

$\Rightarrow -x^2/2 + 8/3(x-1) = 0.$

$\Rightarrow x^2 = 16/3(x-1)$

$\Rightarrow x^2 - 16/3\,x + 16/3 = 0.$

$x = 1.335$ m or 4 m.

The SFD and BMD can be drawn by joining the ordinates at salient point by straight line or smooth curves, as defined by governing equations.

3. Let us draw SFD and BMD diagram for the following figure.

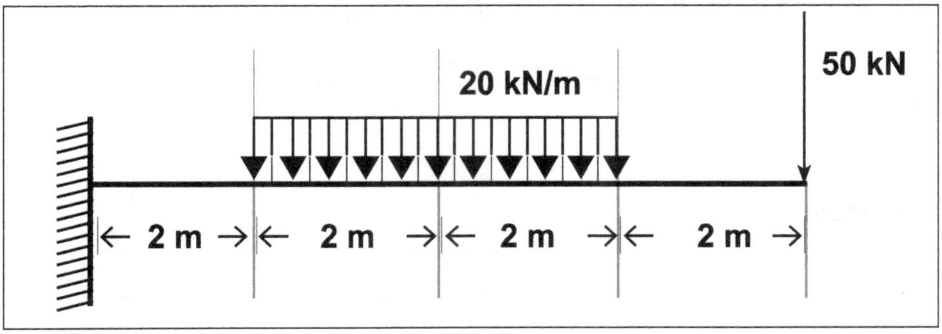

Solution:

At first we shall find the reaction of the cantilever beam.

A cantilever beam is a common type of beam that is supported on a single fixed joint at one end. A fixed joint can provide a vertical reaction, a horizontal reaction and a reaction moment. Thus, while finding reaction we must transform a distributive load (UVL, UDL) to their equivalent concentrated or the point load. An equivalent load of a distributed load shall be found by placing total load at the centroid of the distributed load diagram.

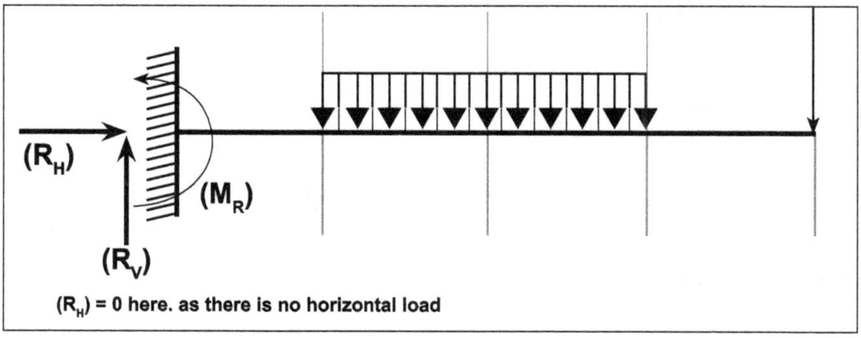

Free Body Diagram of the Beam:

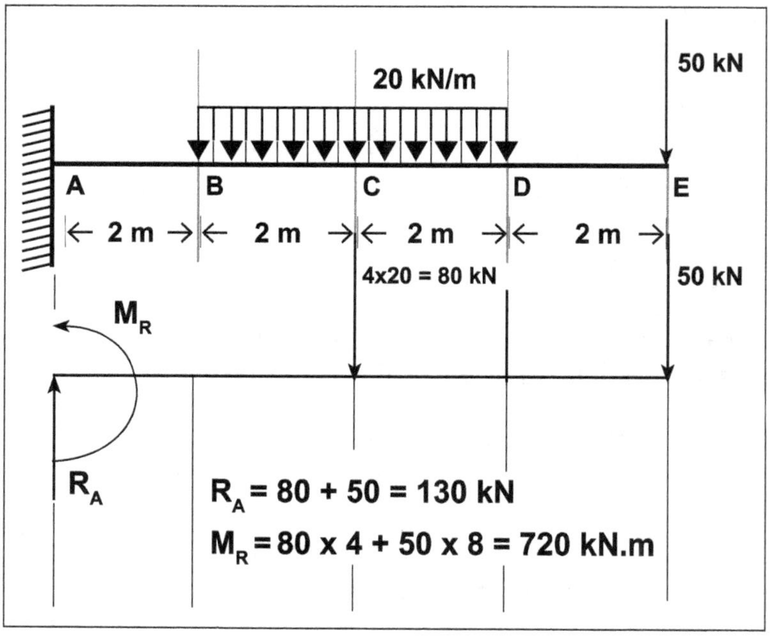

SF and BM Equations:

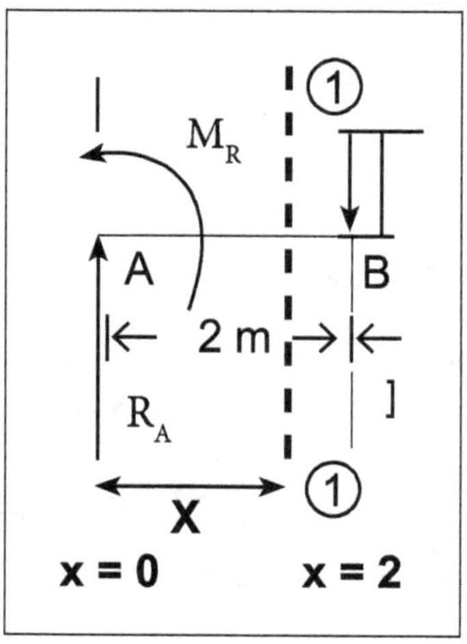

Section AB ($0 \le X \le 2$):

$$SF = R_A = 130 \text{ KN.}$$

$$BM = -M_R + R_A X = -720 + 130X \text{ KN.m.}$$

At, $X = 0$; $SF = 130$ kNand $BM = -720$ KN.m.

At, $X = 2$; $SF = 130$ KN and $BM = -720 + 260 = -460$ KN.m.

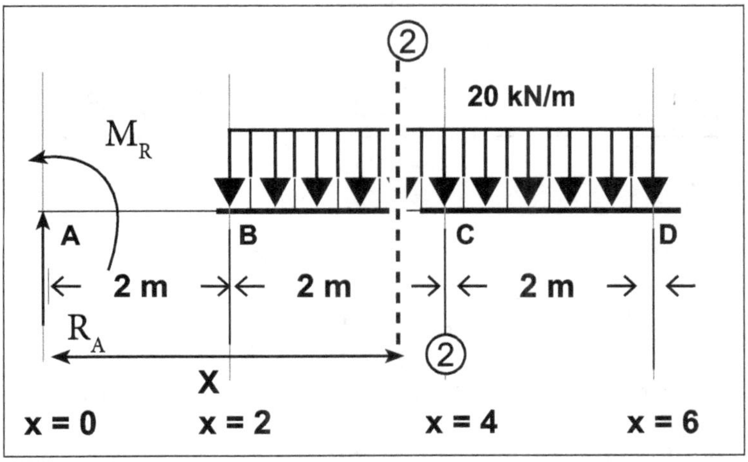

Section BD ($2 \le X \le 6$):

$$SF = R_A - 20(X-2) = 130 - 20(X-2).$$

$$BM = -M_R + R_A X - \{20(X-2)^2\}/2.$$

$$= -720 + 130X - \{20(X-2)^2\}/2$$

At, $X = 2$; $SF = 130$ kN and $BM = -460$ kN.m.

At, $X = 64$; $SF = 130 - 80 = 50$ kN and $BM = -720 + 780 - 160 = -100$ kN.m.

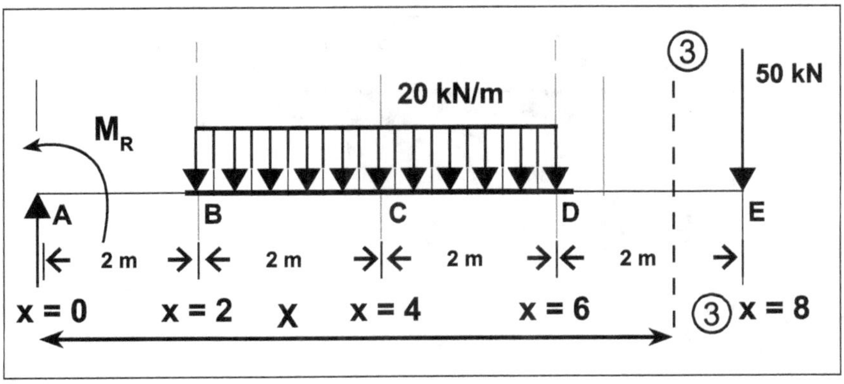

When a distributive load remains fully on the left side of the section line as it is in the above diagram, we should use an equivalent point load in the place of Distributive load of UVL and UDL.

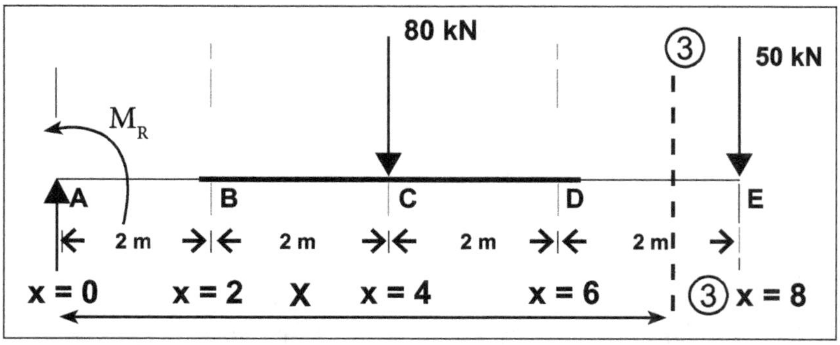

Section DE ($6 \leq X \leq 8$):

$$SF = RA - 80 = 130 - 80 = 50 \text{ KN}.$$

$$BM = -MR + RAX - 80(X-4) = -720 + 130X - 80(X-4).$$

At, $X = 6$; $SF = 130 - 80 = 50$ kN and $BM = -720 + 780 - 160 = -100$ kN.m.

At, $X = 8$; $SF = 130 - 80 = 50$ kN and $BM = -720 + 1040 - 320 = 0$ kN.m.

SFD and BMD diagram:

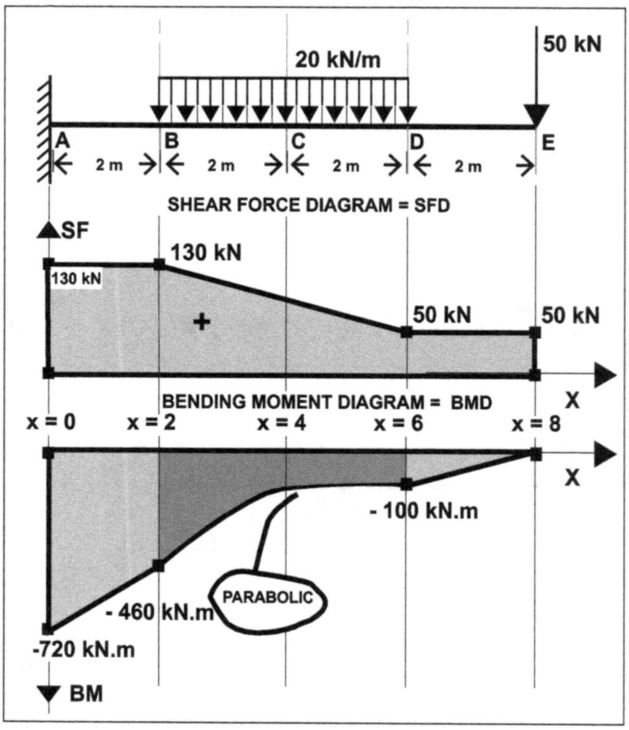

Permissions

All chapters in this book are published with permission under the Creative Commons Attribution Share Alike License or equivalent. Every chapter published in this book has been scrutinized by our experts. Their significance has been extensively debated. The topics covered herein carry significant information for a comprehensive understanding. They may even be implemented as practical applications or may be referred to as a beginning point for further studies.

We would like to thank the editorial team for lending their expertise to make the book truly unique. They have played a crucial role in the development of this book. Without their invaluable contributions this book wouldn't have been possible. They have made vital efforts to compile up to date information on the varied aspects of this subject to make this book a valuable addition to the collection of many professionals and students.

This book was conceptualized with the vision of imparting up-to-date and integrated information in this field. To ensure the same, a matchless editorial board was set up. Every individual on the board went through rigorous rounds of assessment to prove their worth. After which they invested a large part of their time researching and compiling the most relevant data for our readers.

The editorial board has been involved in producing this book since its inception. They have spent rigorous hours researching and exploring the diverse topics which have resulted in the successful publishing of this book. They have passed on their knowledge of decades through this book. To expedite this challenging task, the publisher supported the team at every step. A small team of assistant editors was also appointed to further simplify the editing procedure and attain best results for the readers.

Apart from the editorial board, the designing team has also invested a significant amount of their time in understanding the subject and creating the most relevant covers. They scrutinized every image to scout for the most suitable representation of the subject and create an appropriate cover for the book.

The publishing team has been an ardent support to the editorial, designing and production team. Their endless efforts to recruit the best for this project, has resulted in the accomplishment of this book. They are a veteran in the field of academics and their pool of knowledge is as vast as their experience in printing. Their expertise and guidance has proved useful at every step. Their uncompromising quality standards have made this book an exceptional effort. Their encouragement from time to time has been an inspiration for everyone.

The publisher and the editorial board hope that this book will prove to be a valuable piece of knowledge for students, practitioners and scholars across the globe.

Index

www.ingramcontent.com/pod-product-compliance
Lightning Source LLC
LaVergne TN
LVHW080604210926
848485LV00003B/103